For Grade

3

Science on Target

Parent/Teacher Edition

Using Graphic Organizers to Improve Science Skills

Written By:

Andrea Karch Balas, Ph.D

Published By:
Show What You Know® Publishing
A Division of Englefield & Associates, Inc.
P.O. Box 341348
Columbus, OH 43234-1348
1-877-PASSING (727-7464)

www.showwhatyouknowpublishing.com

Printed in the United States of America
11 10 20 19 18 17 16 15 14 13 12 11 10 9 8 7 6 5 4 3 2 1

ISBN: 1-59230-327-7

About the Author

Andrea Karch Balas, Ph.D., is an educator and a scientist who has taught both in the traditional classroom and in nonformal educational settings, from kindergarten to adult. Andrea has presented her research on the teaching and learning of Science both nationally and internationally. Andrea received her doctorate in education from The Ohio State University. In addition to this book, Andrea is the co-author of the Ready, Set, Show What You Know® series for grades K–3 in Ohio and Florida.

Acknowledgements

Show What You Know® Publishing acknowledges the following for their efforts in making these skill-building materials available for students, parents, and teachers.

Cindi Englefield, President/Publisher
Eloise Boehm-Sasala, Vice President/Managing Editor
Jill Borish, Production Editor
Jennifer Harney, Illustrator/Cover Designer

Table of Contents

Introduction

The Nature of Science in the Classroom

The challenge of teaching and learning Science in the classroom is embedded in the nature of the process and products of Science. The discipline of Science provides the organization for the framework of academic content and the methodology for explorations and experimentation. But classroom Science is a multidisciplinary endeavor; it incorporates all curriculum areas into the acquisition of knowledge. Some of the needed contributing skills include:

Language Arts

- Reading—developing vocabulary, making inferences, interpreting ideas,
- Writing—creating clear descriptions and recording observations;

Mathematics

- Using tools and taking measurements,
- Using the appropriate mathematical processes (addition, subtraction, multiplication, division),
- Using units conversion to metric,
- Graphing and creating data displays;

Communication

- Reading the work of others,
- Writing for others to read,
- Effectively presenting data such as when to best use specific graphing and other informational displays;

Social Studies

- Understanding the impact of the cultural climate on scientific theory building, inventions, and discoveries,
- Discussion about where, when, and how Science happens,
- Learning about who "does" Science.

As teachers meet the challenge of the teaching and learning of Science in the classroom, graphic organizers can provide a support system for students to build skills and comprehension while visually organizing their knowledge.

Graphic Organizer's Use in the Classroom

A graphic organizer is a tool that serves students and teachers. Graphic organizers provide students with a framework to create a visual depiction of their knowledge or thought process. For the teacher, the graphic organizer provides a valuable assessment and evaluation tool. There are many kinds of graphic organizers. (A discussion and depiction of several of them are included later in the text.) For students and teachers, the graphic organizers assist in communicating information about the knowledge of patterns and relationships, connections, and visual presentation of an idea or results of an exploration.

When a graphic organizer is presented for use in the classroom, it is important to model how the tool is used by the students with the presented material. Students should be provided with opportunities to use the graphic organizer in a whole class or group situation before they are expected to independently use the tool. Students should also understand why a particular graphic organizer is selected for use with a particular lesson, assignment, or assessment.

Why Use Graphic Organizers with Your Students?

Graphic organizers can help motivate students because they provide unique ways to think about, arrange, and represent their ideas.

Graphic organizers are easy for students to use for several reasons. These benefits include the format, idea clarification, and visualization.

Format:

- Graphic organizers provide an available format for use to communicate ideas.
- Complex and dense Science topics can be grouped into manageable pieces.
- With the use of words and phrases in the format, connections and relationships are achieved.

Idea clarification:

- It is easy to rework thought processes and rearrange the information on the organizer to clarify thought processes.
- On specific maps, positions and points of view are outlined and supported.
- Manipulation of words and phrases helps to review and demonstrate knowledge and understanding of learned concepts.

Visualization:

- It's easy to edit, revise, and quickly add to a visual map.
- Graphic organizers are tools that can be used for planning the processes for experiments, exploration, and research.
- Graphic organizers provide a visual depiction of concepts and relationships.
- Students with written language and writing skill challenges may find the graphic organizers an easy tool to demonstrate their knowledge.

When to Use Graphic Organizers

Teachers can use graphic organizers during all aspects of the learning process.

As a new topic is introduced, graphic organizers can be used to:

- Discover the students' prior knowledge about a new topic or process,
- Link the new material to previous topics, and
- Identify any student misconceptions about the topic.

During the learning process, graphic organizers can be used to:

- Engage students in active learning by adding and rearranging key ideas, concepts, and processes,
- Appeal to various learning styles with unique visual ways of structuring content material and processes, and
- Provide memory tools to assist students in remembering or understanding concepts, and processes.

Assessing the learning process, graphic organizers can be used to:

- Evaluate student knowledge of connections and relationships,
- Identify student knowledge of the content and concepts, and
- Identify student misconceptions.

Evaluation of Student Work

When scoring the graphic organizers, rubrics are commonly used. Many rubrics are available for use by classroom teachers. The goal of using rubrics is to assess the students' problem-solving abilities and ability to support their answers, processes, or conclusions. Generally with a rubric, point values are assessed for the students' work involving the concept, process, or procedure that is the focus of the lesson. Students are often given the rubric in advance, so they will know how they are going to be evaluated.

When assessing the students' work, it is not necessary to evaluate all aspects of the work each time. By focusing on one or two areas, the students are provided with feedback and possible ways to address specific problem areas. With the point score the teacher might add a comment to stretch the students' thinking. These comments might include:

- Represent this concept, process, or procedure in another way,
- Take another view point on this topic and support it,
- Propose another solution to this issue, and
- Use another tool to represent this concept, process, or procedure.

Sample of a 4-Point Rubric

4 Points

- Contains an effective solution using correct reasoning and/or problem set-up
- Shows complete understanding of the concept, process, or procedure
- Explains all relevant components of the concept, process, or procedure
- Demonstrates logical reasoning and valid conclusions
- Communicates knowledge effectively using words, pictures, and/or symbols

3 Points

- Contains an effective solution with minor errors in reasoning and/or problem set-up
- Shows an understanding of the concept, process, or procedure
- Explains most of the points relative to the solution but not all
- Demonstrates generally reasonable and valid conclusions
- Communicates adequately using words, pictures, and/or symbols

2 Points

- Contains no solution or a flawed solution
- Demonstrates limited understanding of the concept, process, or procedure
- Does not address the most relevant points of the solution
- Faulty reasoning
- Weak conclusion
- Limited or ineffective communication using words, pictures, and/or symbols

1 Point

- Contains no solution
- Demonstrates no knowledge of the scientific concept
- Little or no reasoning
- No solution
- Invalid communication using words, pictures, and/or symbols

0 Points

- No data
- No attempt

Visual Glossary of Graphic Organizers

A graphic organizer is an instructional tool used to illustrate a student's or class's prior knowledge about a topic. This visual glossary will give you an idea of what the graphic organizers used in the *Science on Target for Grade 3 Student Workbook* look like, as well as the best way for your students to put them to practical use in your everyday Science lessons.

Compare and Contrast Chart

	name 1	name 2
attribute 1		
attribute 2		
attribute 3		

Compare and Contrast Chart uses: Show similarities and differences between two things (people, places, events, ideas, etc.).

Examples include: The comparison of plants and animals. Plants and animals are eukaryotes but plant cells contain additional organelles (cell walls, chloroplasts). Plants make their own food through the process of photosynthesis. Plants are the beginning of the food chain for all animals.

Questions to ask: What are the objects, processes, or procedures being compared? What are the component parts of each? How are they similar? How are they different?

Concept Map

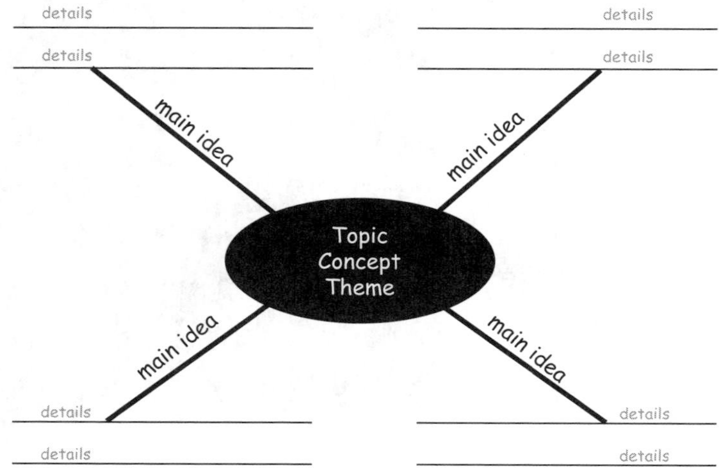

Concept Map uses: Description of a central idea and the relationship of supporting ideas, topics, or functions.

Examples include: If "States of Matter" is the central topic, the connected ideas would be solid state, liquid state, and gaseous state. For each of the states of matter, details could be added to relate the form of the molecules in the state. For example, solids are rigid around a fixed point, and liquids and gases take the shape of their containers.

Questions to ask: How are the ideas connected or interrelated? What details are important to include on the map?

Cycle

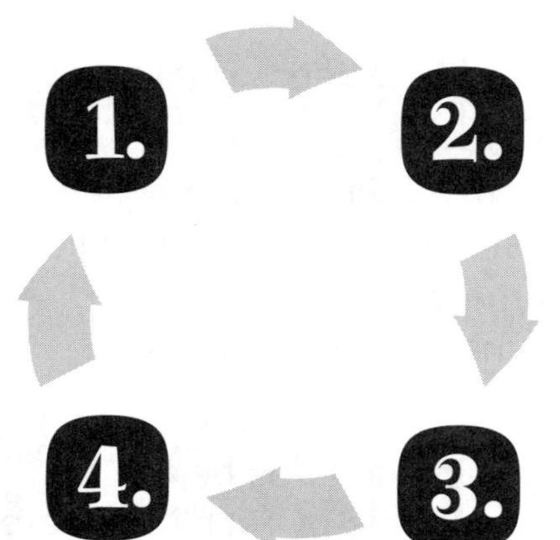

Cycle uses: Illustrates the repeated stages or events that occur to create a specific product or event.

Examples include: The illustration of cell reproduction, rock formation, and weather conditions.

Questions to ask: What are the critical stages or events in the cycle? What is the sequence in the activity? Where or what is the event that promotes the continuity of the cycle?

Diagram

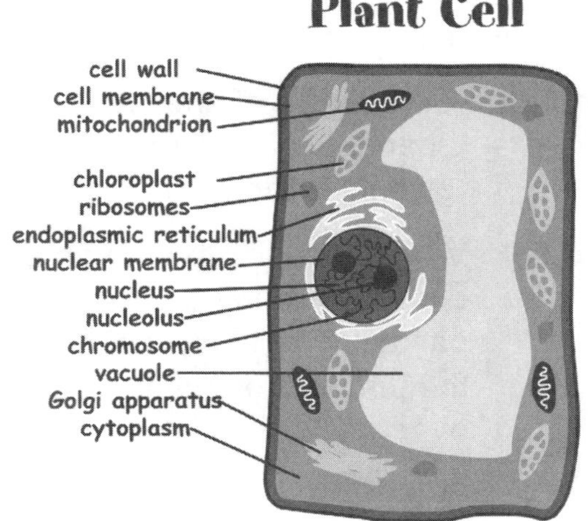

Diagram uses: Provides a tool for describing the relationship between the parts of a system.

Examples include: The model of the cell, the layers of soil, or the components of a circuit.

Questions to ask: What are the parts of the object, theme, or system? What is a visual representation of the information about this process or experiment?

Dichotomous Key

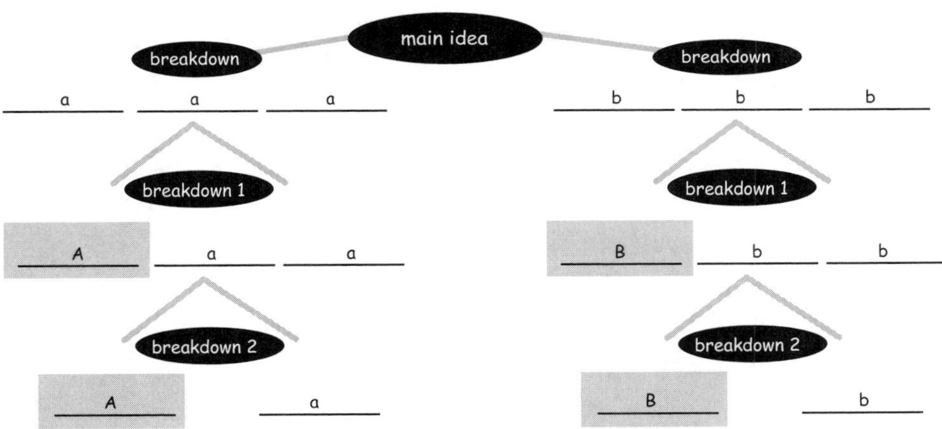

Dichotomous Key uses: Characteristics are used to divide a group of objects, organisms, or ideas into two groups until one object remains in each division.

Examples include: Separating organisms into Kingdoms based on attributes.

Questions to ask: How can I separate these objects, organisms, or ideas based on a given trait?

Graph

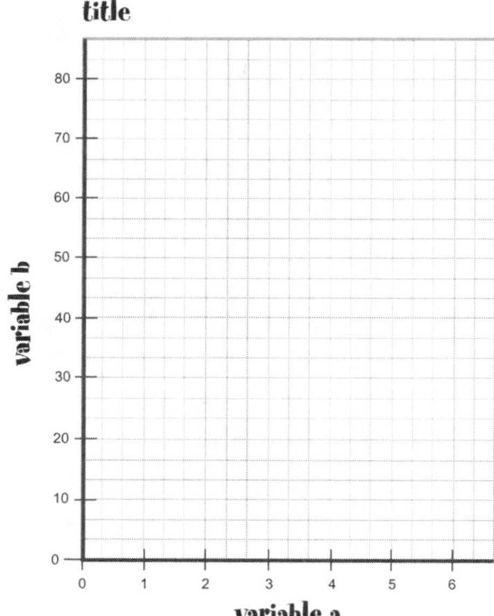

Graph uses: To visually depict collected data.

Examples include: The results of a survey or changes that occur in a given situation.

Questions to ask: Vary depending on the graph, e.g., How many people like a particular fruit? How did the temperature change over time?

Organizational Outline

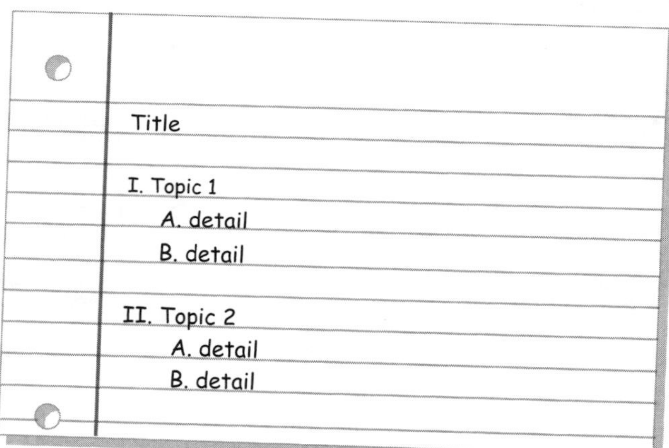

Organizational Outline uses: Organizing information, sequencing processes or events.

Examples include: Organizing information from a Science article or presenting information about a multifaceted topic such as ecosytems.

Questions to ask: What details are given to support the topic?

Series of Events Chain or Flow Chart

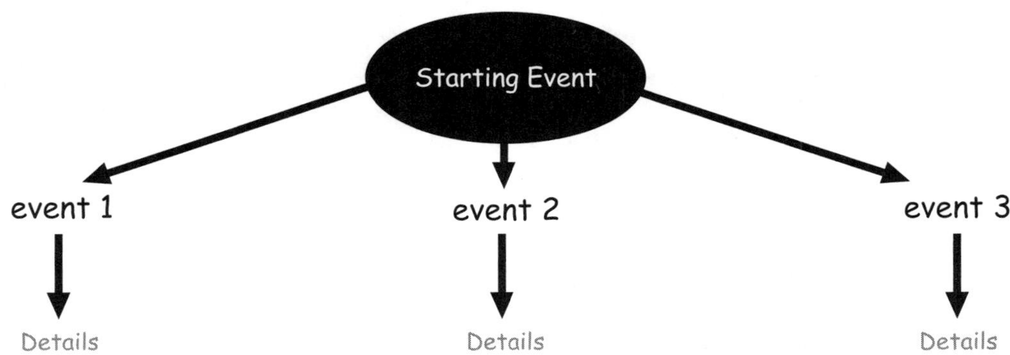

or

Event: ______________________

1. ______________________

then

2. ______________________

then

3. ______________________

then

4. ______________________

then

5. ______________________

Series of Events Chain or Flow Chart uses: Description of stages, steps, sequences, or actions.

Examples include: The formation of something like clouds; the steps in a procedure or experimental design; a sequence of events like erosion or weathering; or the actions leading to the research and development of a product like television or a procedure like space exploration.

Questions to ask: How does the event or process begin? What are the next stages or steps? How are they connected? What is the end product or result?

Timeline

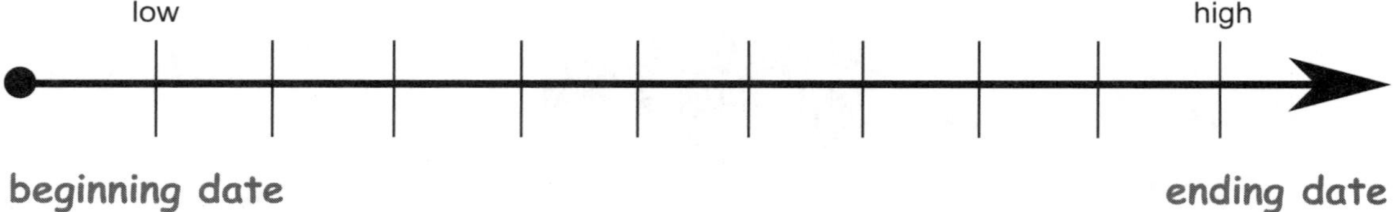

Timeline uses: Showing historical events, creating a record, or charting events.

Examples include: The development of the light bulb, eras of geological time, a growth chart for plants, phases of the moon.

Questions to ask: When does the activity or process begin? When does it end? What happened on a specific date in time?

Venn Diagram

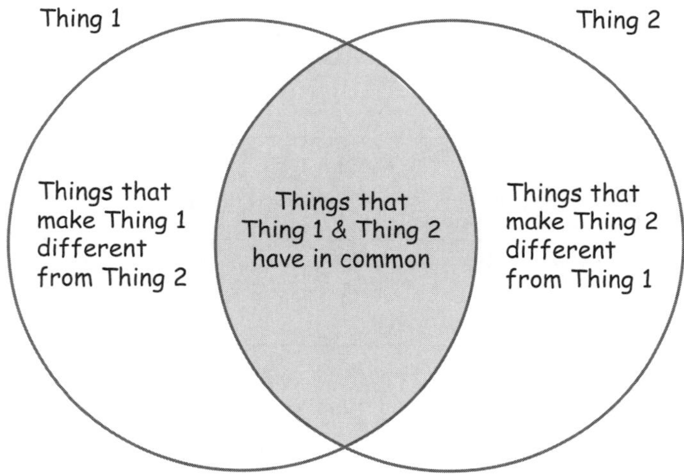

Venn Diagram uses: The comparison and contrast of objects, events, organisms, or themes.

Examples include: The comparison of the microscope and telescope. They both assist the eye in seeing objects. The microscope enlarges smaller objects; the telescope makes objects that are far away closer for observation.

Questions to ask: How are the objects, events, or themes the same? How are they different?

Chapter 1

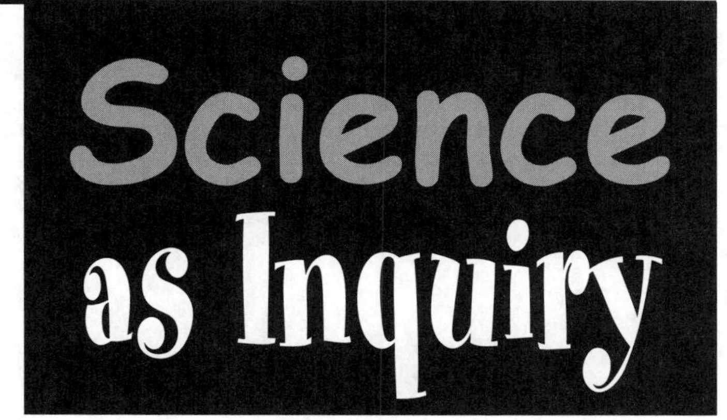

Chapter 1 of the *Science on Target for Grade 3 Student Workbook*, covers the National Content Standards for Science as Inquiry. The standards are as follows:

Science as Inquiry

Abilities necessary to do scientific inquiry

- Ask a question about objects, organisms, and events in the environment.
- Plan and conduct a simple investigation.
- Employ simple equipment and tools to gather data and extend the senses.
- Use data to construct a reasonable explanation.
- Communicate investigations and explanations.

Understanding scientific inquiry

- Ask and answer questions then compare the answer to what is already known.
- Use different types of investigations depending on the questions students are trying to answer, such as describing, classifying, and experimenting.
- Use a variety of simple instruments, such as rulers, magnifiers, and thermometers.
- Explain observations using evidence and what students already know.
- Report the results of investigations to others; describe the investigations so others can repeat them.
- Review the work of others and ask questions about the work of others.

All of the pages from Chapter 1 of the *Science on Target for Grade 3 Student Workbook*, are reproduced in this Parent/Teacher Edition in reduced-page format with sample answers. These activities will help your students develop the skills necessary to do scientific inquiry and understand scientific inquiry.

Students should use the “Clues for Success” Checklist, for each activity in this section, as a tool to help them do their best work.

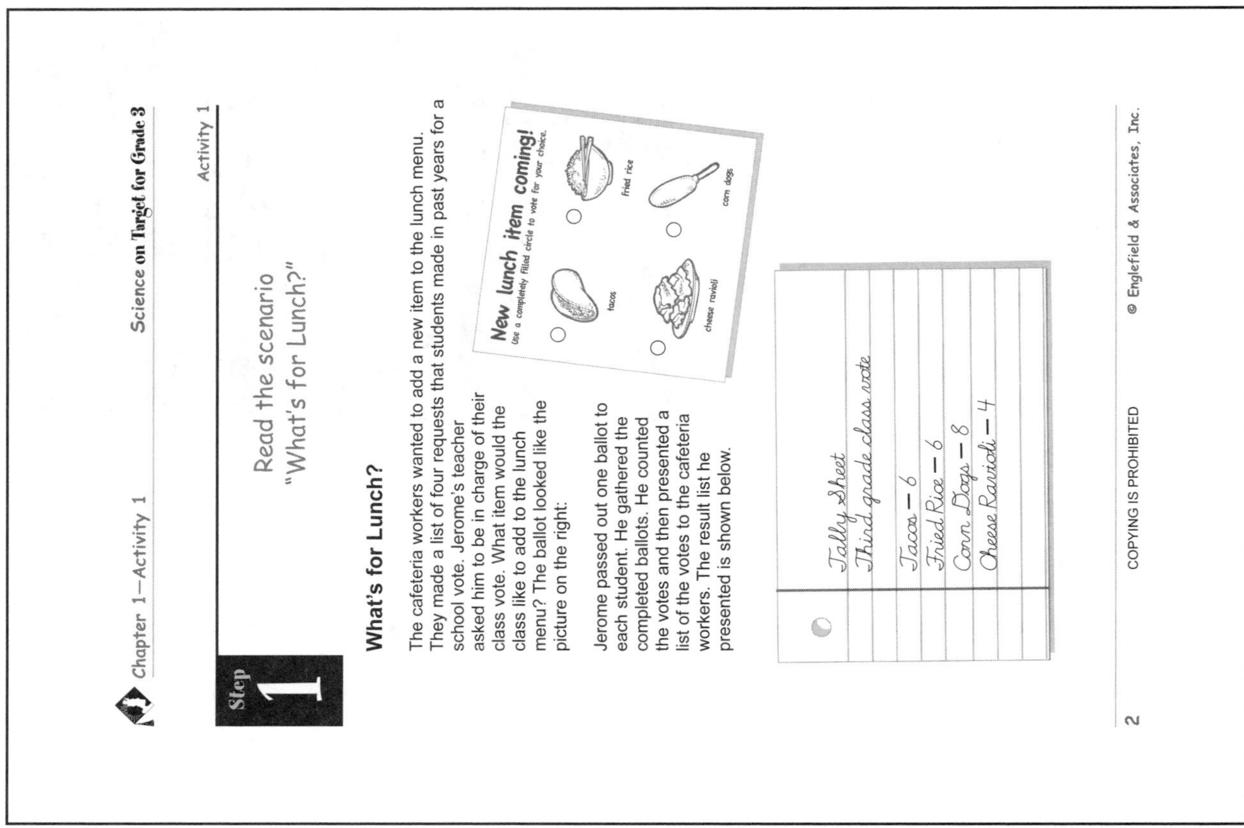

Step 1

Read the scenario "What's for Lunch?"

What's for Lunch?

The cafeteria workers wanted to add a new item to the lunch menu. They made a list of four requests that students made in past years for a school vote. Jerome's teacher asked him to be in charge of their class vote. What item would the class like to add to the lunch menu? The ballot looked like the picture on the right:

Jerome passed out one ballot to each student. He gathered the completed ballots. He counted the votes and then presented a list of the votes to the cafeteria workers. The result list he presented is shown below.

Note: Student answers may vary. Example responses are for use as a guide.

Chapter 1

Science as Inquiry

The activities in this section of the book will focus on Science as Inquiry.

These activities will help you develop the skills necessary to do scientific inquiry and understand scientific inquiry. These activities include:

- asking questions about the world around you,
- planning simple investigations,
- selecting and using scientific tools and equipment,
- using data to explain your observations or experiment results,
- communicating your ideas and results, and
- reviewing the work of other groups.

Use the "Clues for Success" Checklist, as you complete each activity in this section, as a tool to help you do your best work.

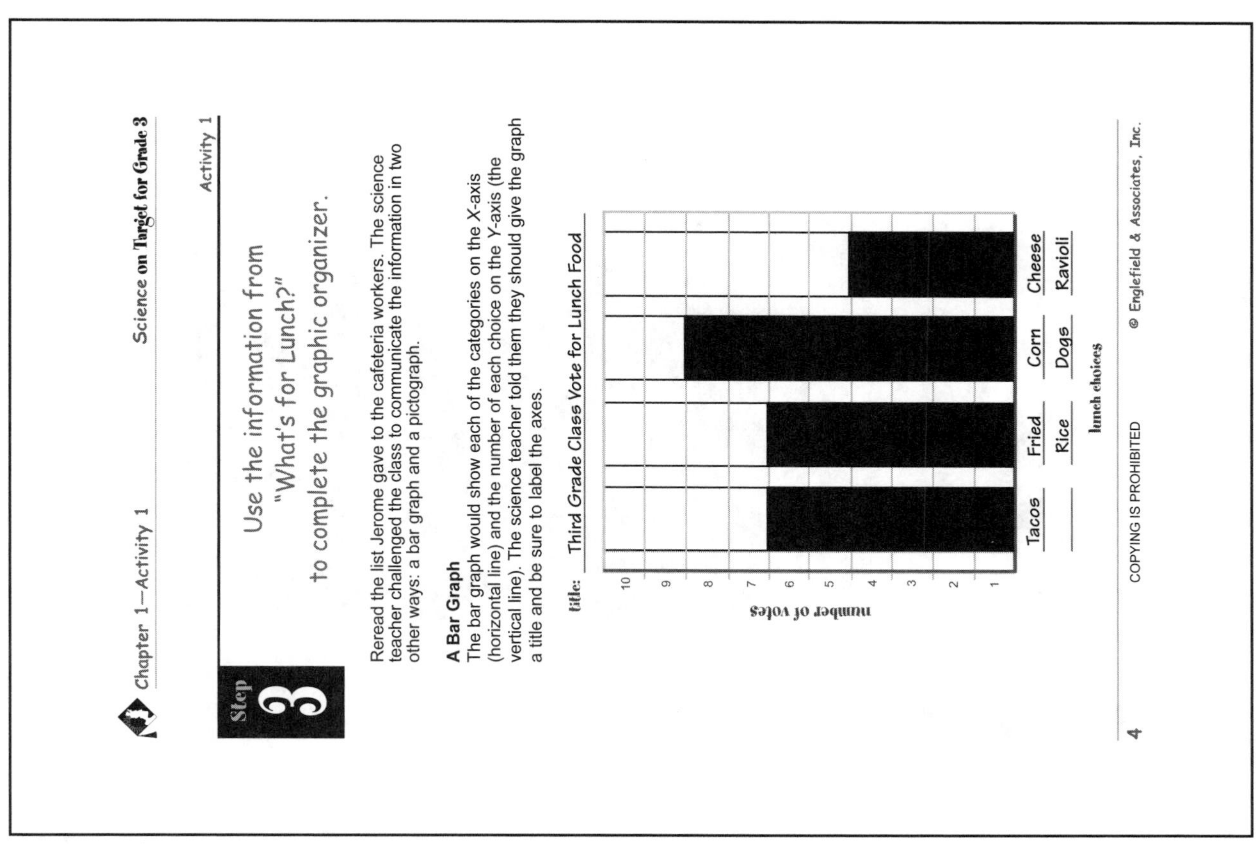

Chapter 1—Activity 1 Science on Target for Grade 3

Activity 1

Step 3 Use the information from "What's for Lunch?" to complete the graphic organizer.

Reread the list Jerome gave to the cafeteria workers. The science teacher challenged the class to communicate the information in two other ways: a bar graph and a pictograph.

A Bar Graph

The bar graph would show each of the categories on the *X*-axis (horizontal line) and the number of each choice on the *Y*-axis (the vertical line). The science teacher told them they should give the graph a title and be sure to label the axes.

4 COPYING IS PROHIBITED © Englefield & Associates, Inc.

Note: Student answers may vary. Example responses are for use as a guide.

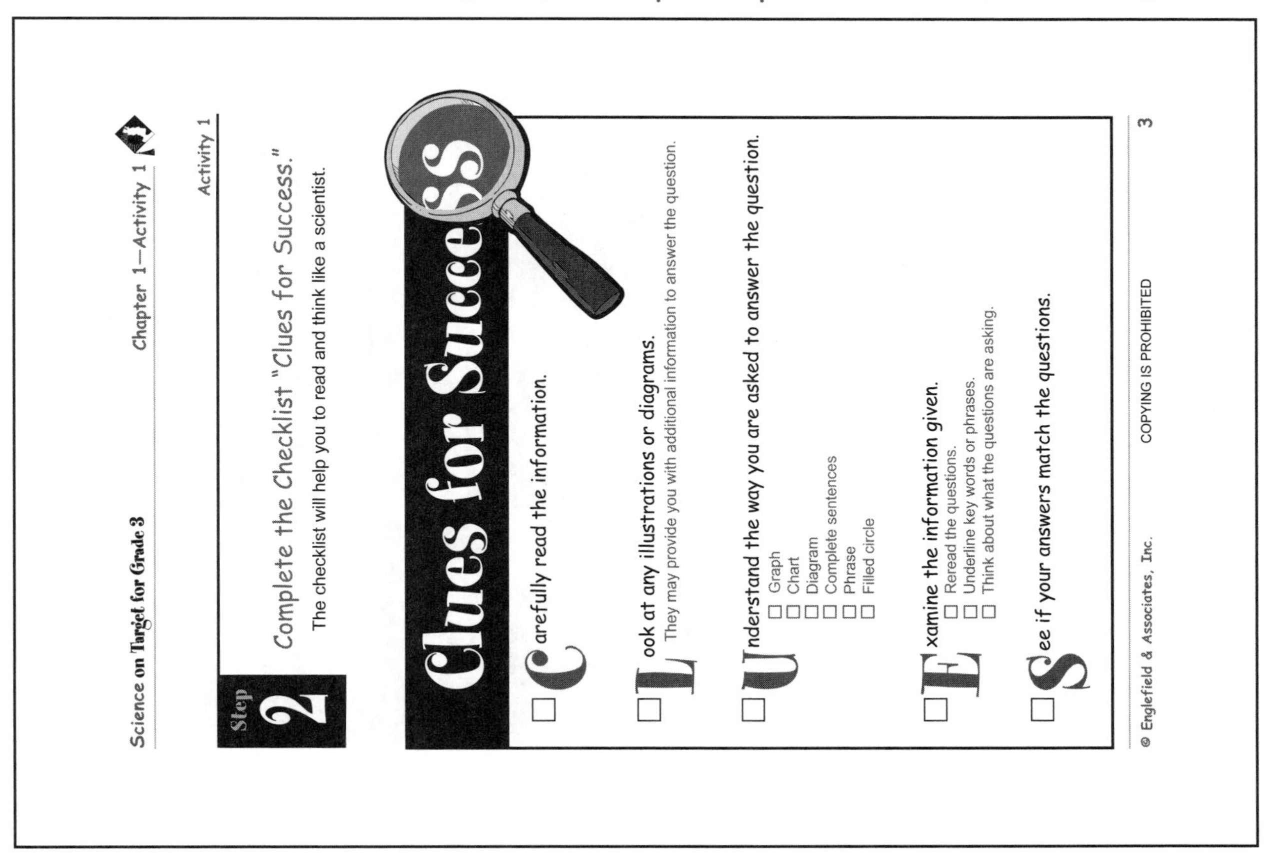

Science on Target for Grade 3 Chapter 1—Activity 1

Activity 1

Step 2 Complete the Checklist "Clues for Success."

The checklist will help you to read and think like a scientist.

Clues for Success

- ☐ **C**arefully read the information.
- ☐ **L**ook at any illustrations or diagrams.
  They may provide you with additional information to answer the question.
- ☐ **U**nderstand the way you are asked to answer the question.
  - ☐ Graph
  - ☐ Chart
  - ☐ Diagram
  - ☐ Complete sentences
  - ☐ Phrase
  - ☐ Filled circle
- ☐ **E**xamine the information given.
  - ☐ Reread the questions.
  - ☐ Underline key words or phrases.
  - ☐ Think about what the questions are asking.
- ☐ **S**ee if your answers match the questions.

© Englefield & Associates, Inc. COPYING IS PROHIBITED 3

Step 4

Answer the following questions for "What's for Lunch?" using information from your graphic organizer.

When Jerome returned from presenting the class vote to the cafeteria workers, the science teacher complemented his scientific abilities. Use an example from the scenario to show how Jerome demonstrated the scientific abilities listed on this classroom poster.

Note: Student answers may vary. Example responses are for use as a guide.

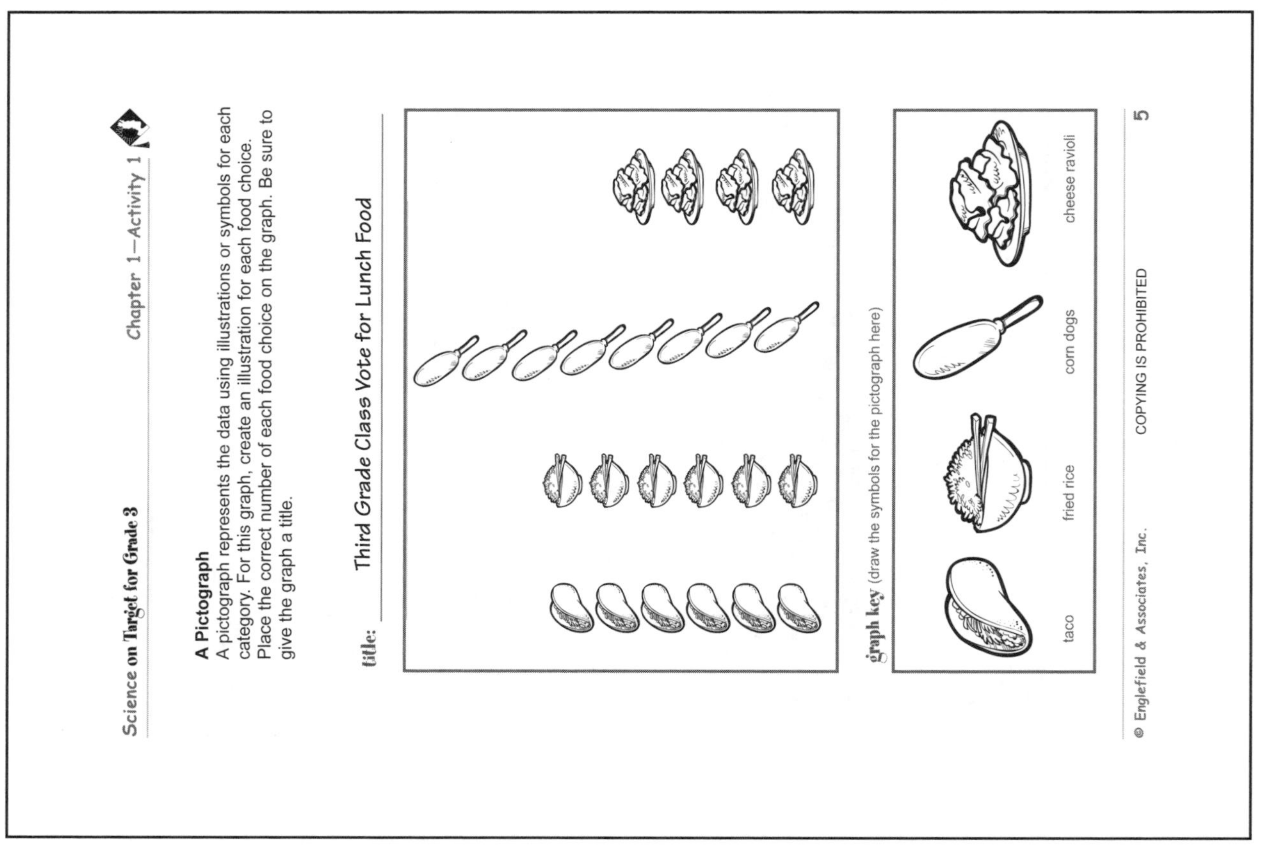

A Pictograph

A pictograph represents the data using illustrations or symbols for each category. For this graph, create an illustration for each food choice. Place the correct number of each food choice on the graph. Be sure to give the graph a title.

title: *Third Grade Class Vote for Lunch Food*

graph key (draw the symbols for the pictograph here)

1. Asks questions. Jerome asked the students to select their favorite food item.

2. Plans and conducts investigations. Jerome made a ballot with all of the food choices on it. He passed the ballots out for a class vote.

3. Uses simple equipment and tools to gather data. Jerome used a pencil and paper ballot to gather data.

4. Uses the data to construct a reasonable result. Jerome used the ballots to add up the votes. He made a tally sheet with results for the cafeteria workers.

5. Clearly communicates the result. Jerome made a results sheet for the cafeteria workers.

Note: Student answers may vary. Example responses are for use as a guide.

Activity 2

Step 1

Read the scenario
"The Class Does Science."

The Class Does Science

The teacher explained that during the school year the class would be conducting different kinds of scientific investigations based on what they want to learn and the questions they ask. He told them about each type of investigation and gave examples of each.

STUDY GUIDE

Describing uses the senses or tools to give details about an event, an object or organism, or a reaction.
Examples: the length of a hallway, the sound of sandpaper on wood, and the feel of a rough-textured fabric.

Classifying uses one or more characteristics of events, objects or organisms, or reactions to divide them into groups.
Examples: sorting crayons by color, grouping organisms into kingdoms, and separating leaves by type.

Experimenting is a series of events that may begin with a question or observation and is followed by an investigation. Observations are made using the senses and/or tools. The process and observations are recorded, then communicated and reviewed by others.
Examples: learning what happens to ice in the sun; finding out how two liquids act when they come together, and asking what happens when plants are watered with a soft drink.

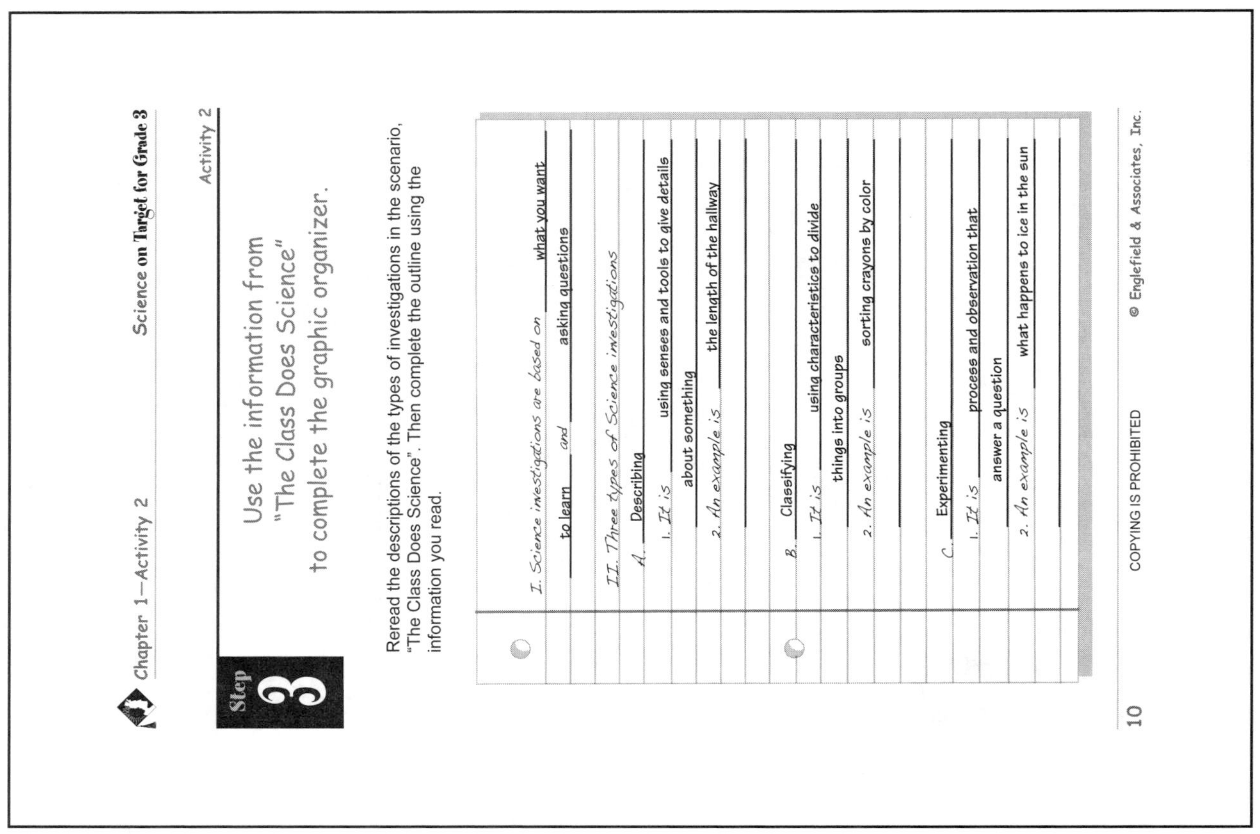
Science on Target for Grade 3 — Chapter 1—Activity 2 — Activity 2

Step 3

Use the information from "The Class Does Science" to complete the graphic organizer.

Reread the descriptions of the types of investigations in the scenario, "The Class Does Science". Then complete the outline using the information you read.

I. Science investigations are based on what you want to learn and asking questions

II. Three types of Science investigations

A. Describing

1. It is using senses and tools to give details about something
2. An example is the length of the hallway

B. Classifying

1. It is using characteristics to divide things into groups
2. An example is sorting crayons by color

C. Experimenting

1. It is process and observation that answer a question
2. An example is what happens to ice in the sun

10 COPYING IS PROHIBITED © Englefield & Associates, Inc.

Note: Student answers may vary. Example responses are for use as a guide.

Science on Target for Grade 3 — Chapter 1—Activity 2 — Activity 2

Step 2

Complete the Checklist "Clues for Success."

The checklist will help you to read and think like a scientist.

Clues for Success

- ☐ **C**arefully read the information.
- ☐ **L**ook at any illustrations or diagrams. They may provide you with additional information to answer the question.
- ☐ **U**nderstand the way you are asked to answer the question.
  - ☐ Graph
  - ☐ Chart
  - ☐ Diagram
  - ☐ Complete sentences
  - ☐ Phrase
  - ☐ Filled circle
- ☐ **E**xamine the information given.
  - ☐ Reread the questions.
  - ☐ Underline key words or phrases.
  - ☐ Think about what the questions are asking.
- ☐ **S**ee if your answers match the questions.

© Englefield & Associates, Inc. COPYING IS PROHIBITED 9

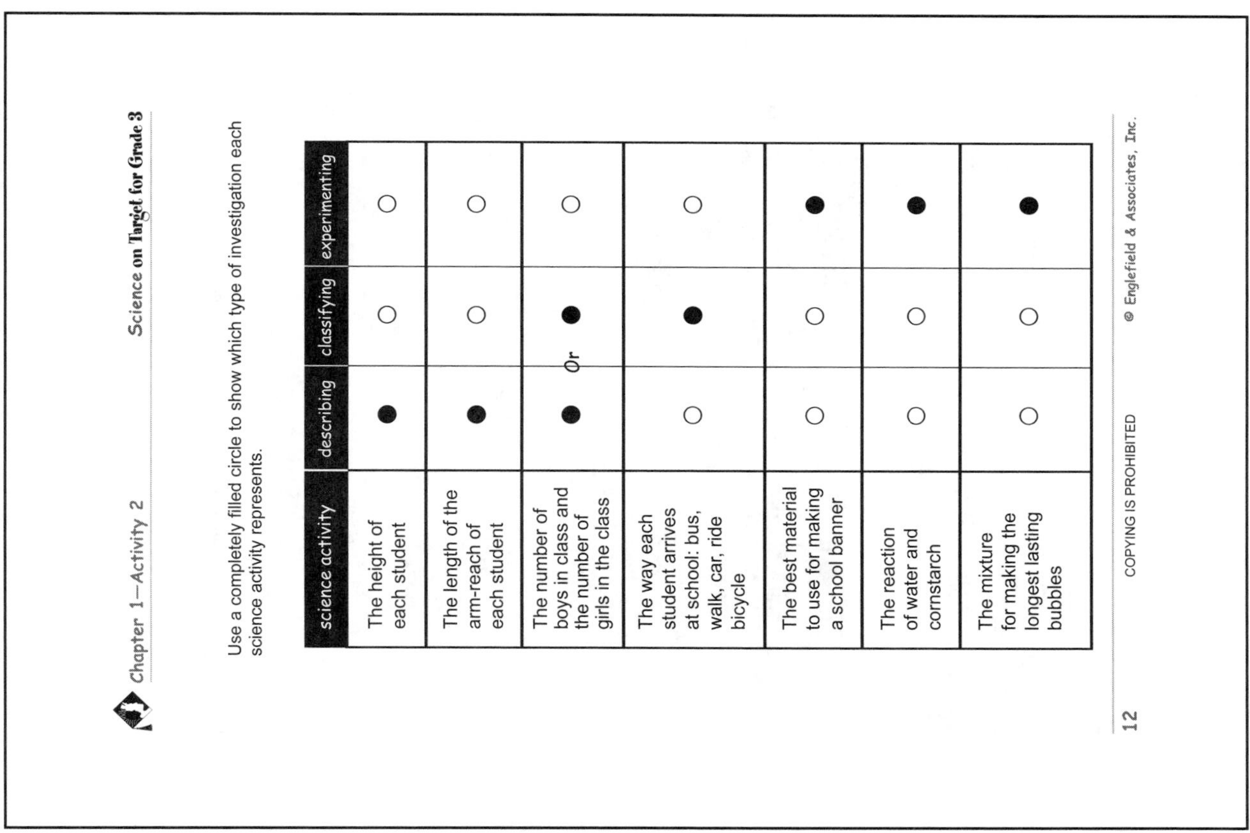

Use a completely filled circle to show which type of investigation each science activity represents.

science activity	describing	classifying	experimenting
The height of each student	●	○	○
The length of the arm-reach of each student	●	○	○
The number of boys in class and the number of girls in the class	●	or ●	○
The way each student arrives at school: bus, walk, car, ride bicycle	○	●	○
The best material to use for making a school banner	○	○	●
The reaction of water and cornstarch	○	○	●
The mixture for making the longest lasting bubbles	○	○	●

Note: Student answers may vary. Example responses are for use as a guide.

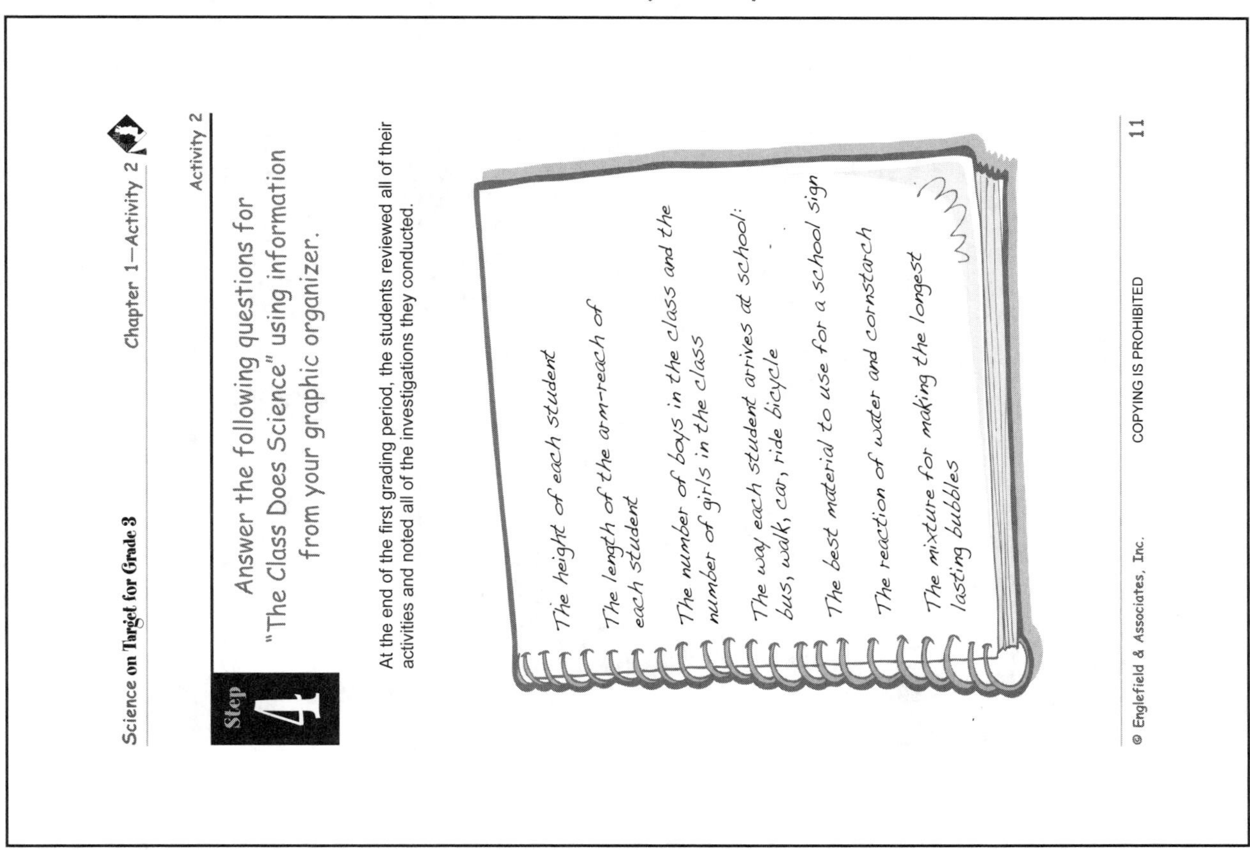

Step 4

Activity 2

Answer the following questions for "The Class Does Science" using information from your graphic organizer.

At the end of the first grading period, the students reviewed all of their activities and noted all of the investigations they conducted.

The height of each student

The length of the arm-reach of each student

The number of boys in the class and the number of girls in the class

The way each student arrives at school: bus, walk, car, ride bicycle

The best material to use for a school sign

The reaction of water and cornstarch

The mixture for making the longest lasting bubbles

Chapter 1—Activity 3 Science on Target for Grade 3

Activity 3

Step 2

Complete the Checklist "Clues for Success."

The checklist will help you to read and think like a scientist.

Clues for Success

☐ Carefully read the information.

☐ Look at any illustrations or diagrams.
They may provide you with additional information to answer the question.

☐ Understand the way you are asked to answer the question.
☐ Graph
☐ Chart
☐ Diagram
☐ Complete sentences
☐ Phrase
☐ Filled circle

☐ Examine the information given.
☐ Reread the questions.
☐ Underline key words or phrases.
☐ Think about what the questions are asking.

☐ See if your answers match the questions.

14 COPYING IS PROHIBITED © Englefield & Associates, Inc.

Note: Student answers may vary. Example responses are for use as a guide.

Science on Target for Grade 3 Chapter 1—Activity 3

Activity 3

Step 1

Read the scenario
"The Car Investigation."

The Car Investigation

The class was conducting an investigation using toy cars to show how adding weight to the cars might change the distance they travel. The teacher stressed that each team should describe what they did so another team could repeat their investigation or ask them questions about their work.

The day before the lab
Each of the teams planned a simple investigation to conduct. To make sure all the equipment they needed would be available, each team gave the team leader a list of materials. Each team requested a meter stick and a roll of tape. Two of the four teams requested a stop watch. The teacher told them she would have a selection of cars and a beaker of washers and pennies available for them to use. Most of the teams wrote a step-by-step plan in their lab journals. Team 3 decided it would be easier to use note cards for their procedure.

The day of the lab
The day of the investigation, each of the teams followed their plans, but Team 3's notecards fell on the floor and had to be placed back into order before they could begin their investigation.

© Englefield & Associates, Inc. COPYING IS PROHIBITED 13

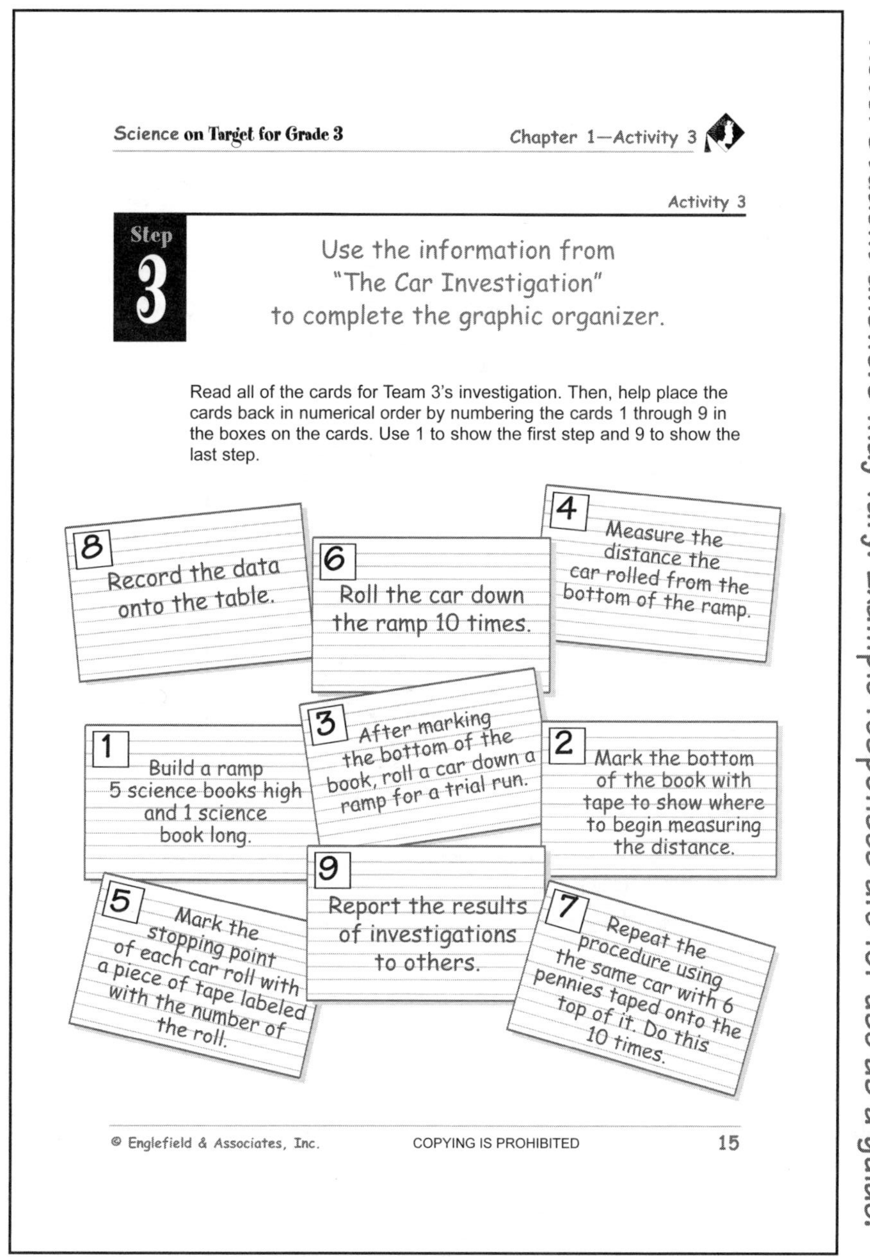

Science on Target for Grade 3 — Chapter 1—Activity 3

Activity 3

Step 3

Use the information from "The Car Investigation" to complete the graphic organizer.

Read all of the cards for Team 3's investigation. Then, help place the cards back in numerical order by numbering the cards 1 through 9 in the boxes on the cards. Use 1 to show the first step and 9 to show the last step.

8 Record the data onto the table.

6 Roll the car down the ramp 10 times.

4 Measure the distance the car rolled from the bottom of the ramp.

1 Build a ramp 5 science books high and 1 science book long.

3 After marking the bottom of the book, roll a car down a ramp for a trial run.

2 Mark the bottom of the book with tape to show where to begin measuring the distance.

5 Mark the stopping point of each car roll with a piece of tape labeled with the number of the roll.

9 Report the results of investigations to others.

7 Repeat the procedure using the same car with 6 pennies taped onto the top of it. Do this 10 times.

© Englefield & Associates, Inc. COPYING IS PROHIBITED 15

Note: Student answers may vary. Example responses are for use as a guide.

Chapter 1—Activity 3 — Science on Target for Grade 3

Activity 3

Step 4

Answer the following questions for "The Car Investigation" using information from your graphic organizer.

1. When you read the steps that Team 3 wrote, do you have enough information to repeat their investigation?

● yes ○ no

Why?

They gave clear directions for the setup. They told the number of trials for each car. They remembered to record and share data.

2. List **two** questions you might ask Team 3 about their work.

1. *What does your data table look like?*

2. *How did you hold and release the car?*

16 COPYING IS PROHIBITED © Englefield & Associates, Inc.

Activity 4

Step 1

Read the scenario "What's in the Bag?"

What's in the Bag?

The class was learning to make observations about the physical properties of objects. The teacher reminded them that their senses were the first tools they could use to describe objects.

The senses include seeing, hearing, smelling, touching, and tasting.

- Seeing could help them describe size and color like the sight of a large orange pumpkin, a small yellow apple, or a blue liquid in a beaker.
- Hearing could help them determine sounds made by or with an object like the sound of a chirping cricket, the rustling of leaves, or the crashing of glass.
- Smelling could identify particular odors like fruit juice, pine trees, or cleaning supplies.
- Touching could reveal texture or weight. Is the surface of the rock smooth or rough? Is the surface spongy or stiff? Is the round object heavier or lighter than a baseball?
- Tasting (if permitted in the investigation) could identify a range of sensations like salty, spicy, sweet, or bitter.

Then, the teacher gave each team a brown lunch bag that was stapled closed. She asked each team to feel the number of objects in the bag and list them **without** opening the bag.

Note: Student answers may vary. Example responses are for use as a guide.

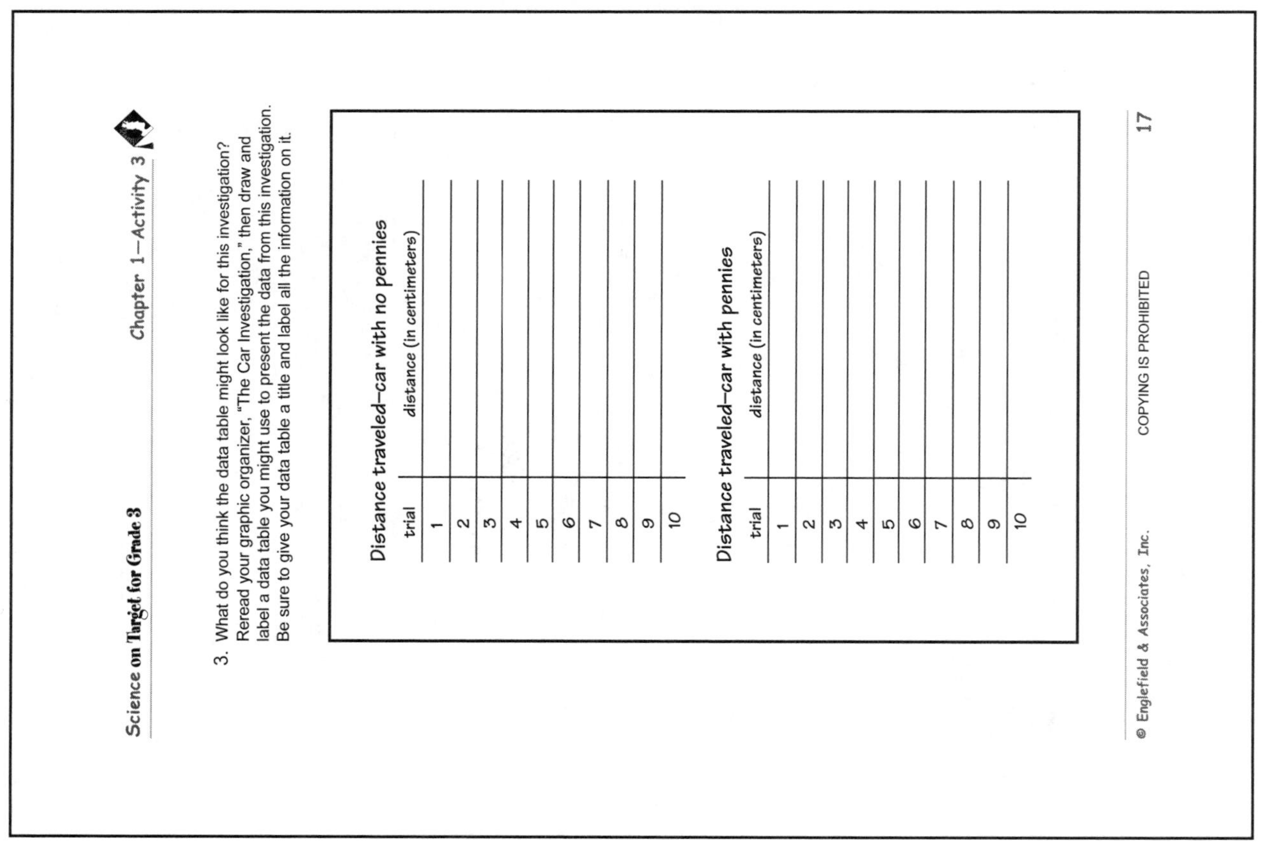

3. What do you think the data table might look like for this investigation? Reread your graphic organizer, "The Car Investigation," then draw and label a data table you might use to present the data from this investigation. Be sure to give your data table a title and label all the information on it.

Distance traveled—car with no pennies

trial	distance (in centimeters)
1	
2	
3	
4	
5	
6	
7	
8	
9	
10	

Distance traveled—car with pennies

trial	distance (in centimeters)
1	
2	
3	
4	
5	
6	
7	
8	
9	
10	

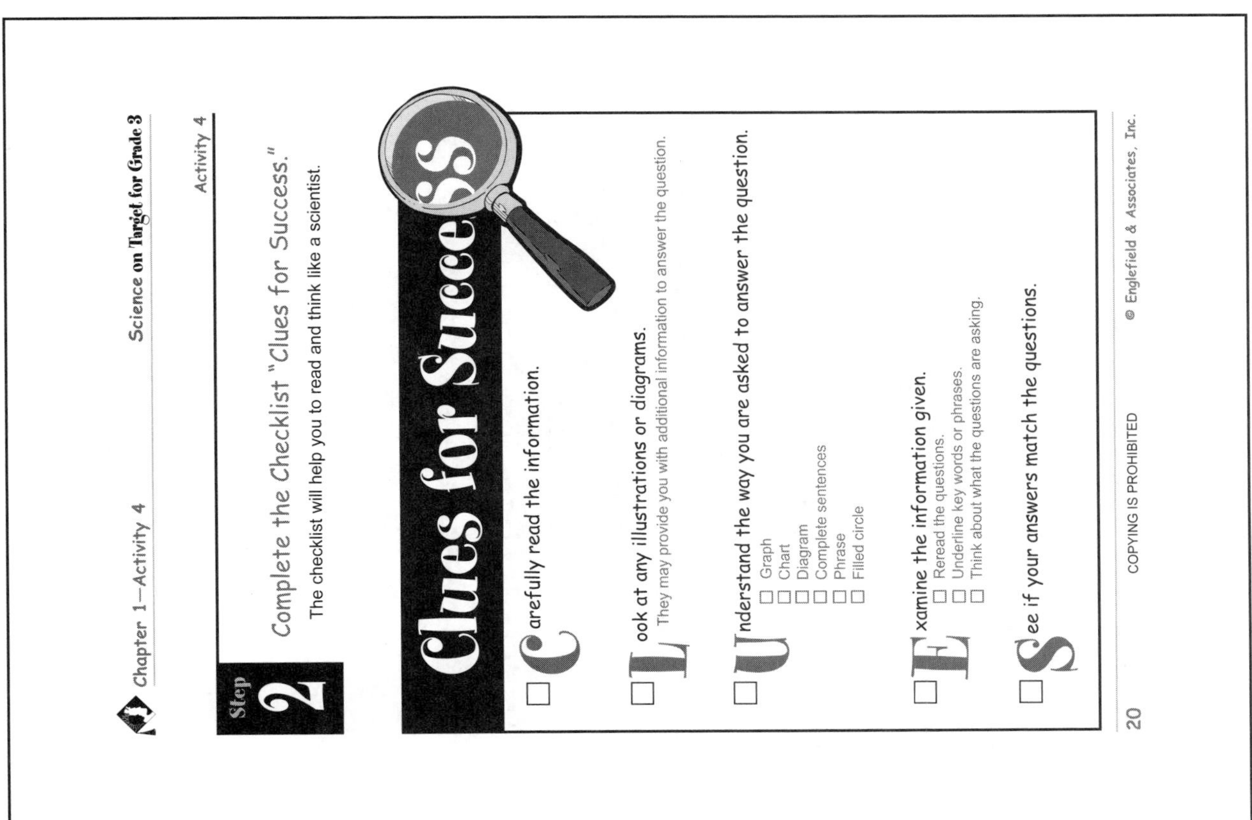

Activity 4

Step 2 Complete the Checklist "Clues for Success."

The checklist will help you to read and think like a scientist.

Clues for Success

- ☐ **C**arefully read the information.
- ☐ **L**ook at any illustrations or diagrams.
  They may provide you with additional information to answer the question.
- ☐ **U**nderstand the way you are asked to answer the question.
  - ☐ Graph
  - ☐ Chart
  - ☐ Diagram
  - ☐ Complete sentences
  - ☐ Phrase
  - ☐ Filled circle
- ☐ **E**xamine the information given.
  - ☐ Reread the questions.
  - ☐ Underline key words or phrases.
  - ☐ Think about what the questions are asking.
- ☐ **S**ee if your answers match the questions.

Note: Student answers may vary. Example responses are for use as a guide.

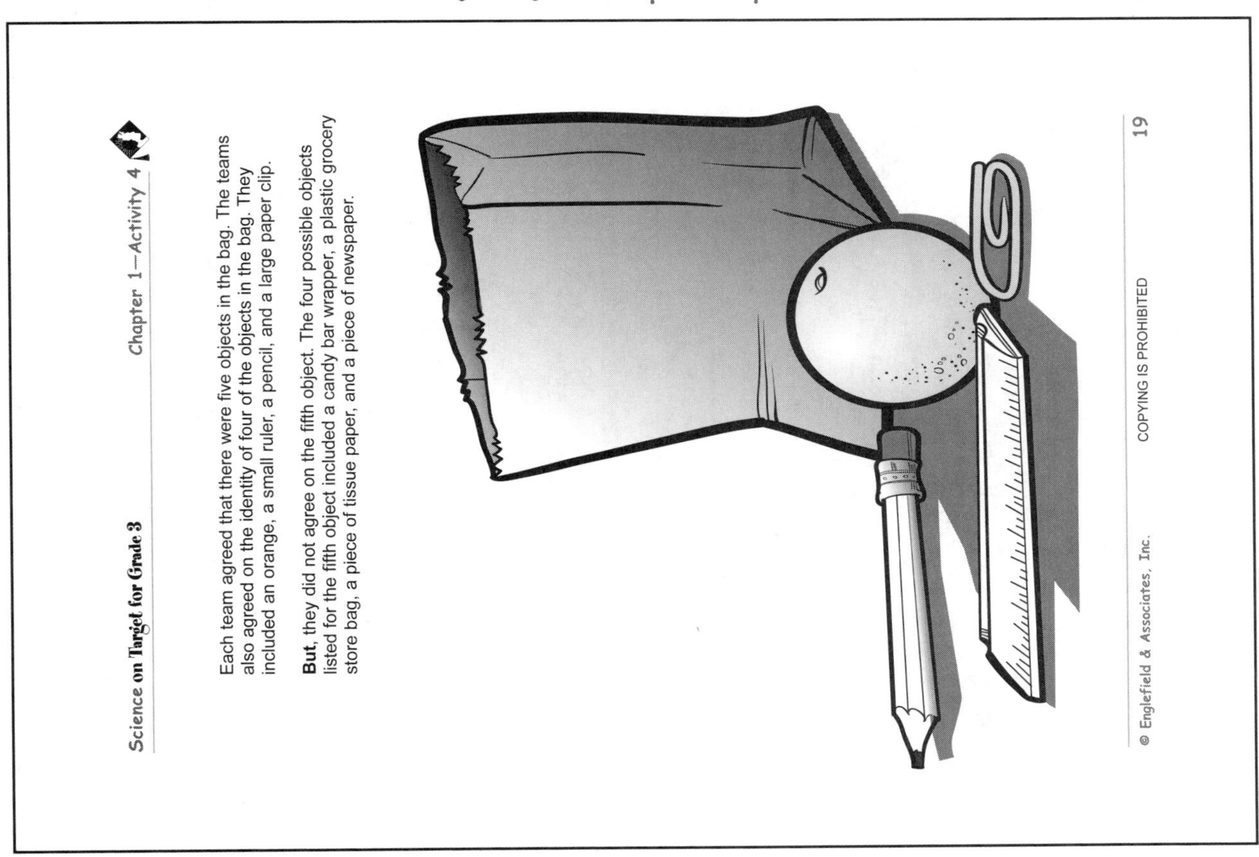

Each team agreed that there were five objects in the bag. The teams also agreed on the identity of four of the objects in the bag. They included an orange, a small ruler, a pencil, and a large paper clip.

But, they did not agree on the fifth object. The four possible objects listed for the fifth object included a candy bar wrapper, a plastic grocery store bag, a piece of tissue paper, and a piece of newspaper.

Activity 4

Step 3

Use the information from "What's in the Bag?" to complete the graphic organizer.

Using the four items that all of the teams agreed on, answer the questions to show which senses were used to help the students identify these objects. (Hint: All senses may not be used for each object.)

What did the students learn from their senses that made them guess an **orange** was in the bag?

Seeing *saw round shape*

Hearing *the sound of the object tapping another object*

Smelling *the scent of an orange*

Touching *felt round object the size of an orange*

What did the students learn from their senses that made them guess a **small ruler** was in the bag?

Seeing *the size of the object in the bag*

Hearing *the sound of the object tapping another object*

Smelling *did not help*

Touching *the flat sides and rectangular shape*

Note: Student answers may vary. Example responses are for use as a guide.

What did the students learn from their senses that made them guess a **pencil** was in the bag?

Seeing *the size of the object*

Hearing *the sound of the object tapping another object*

Smelling *did not help*

Touching *the shape and texture of the object*

What did the students learn from their senses that made them guess a **large paper clip** was in the bag?

Seeing *the size and shape of the object*

Hearing *the sound of the object tapping another object*

Smelling *did not help*

Touching *the shape and texture of the object*

2. All of the teams guessed a small ruler. Use a complete sentence to tell you why the teams did not think the ruler was a regular size ruler (12 inches)?

They did not think a 12-inch ruler would fit in the paper bag.

The 12-inch ruler would be too long!

Note: Student answers may vary. Example responses are for use as a guide.

Activity 4

Step 4

Answer the following questions for "What's in the Bag?" using information from your graphic organizer.

The teams did not agree on the fifth object in the bag. The team ideas included a jumbo size candy bar wrapper, a plastic grocery bag, a piece of tissue paper, and a piece of newspaper.

1. List **two** common physical properties of these objects that could lead the teams to choose these different answers? Identify the sense that helped to identify this property.

   1. *the texture of the object*

      Sense used to help: *touching/feeling*

   2. *the sound it made when touched*

      Sense used to help: *hearing*

Chapter 2

Chapter 2 of the *Science on Target for Grade 3 Student Workbook*, covers the National Content Standards for Physical Science. The standards are as follows:

Physical Science

Properties of objects and materials

- Observe the properties of matter such as size, weight, color, and temperature. Notice how the properties of objects react with those of other objects. Measure the properties of matter using tools, such as rulers, balances, and thermometers.
- Notice that objects made of one or more materials can be described by the properties from which they are made (paper, wood, metal). The properties of matter can be used to separate or sort a group of objects or materials.
- Materials can exist in different states—solid, liquid, or gas. Some materials can be changed by heating or cooling (water).

Position of objects and materials

- The position of an object can be described by locating it relative to another object or background.
- An object's motion can be described by tracing and measuring its position over time.
- An object's position can be changed by pushing or pulling. The size of change is related to the strength of the push or pull.

Light, heat, electricity, and magnetism

- Light travels in a straight line until it strikes an object. Light can be reflected by a mirror, refracted by a lens, or absorbed by an object.
- Heat can be produced in many ways such as burning, rubbing, or mixing one substance with another. Heat can move from one substance to another by conduction.
- Electricity in circuits can produce light, heat, sound, and magnetic effects.

- Electrical circuits require a complete loop through which a current can pass.
- Magnets attract and repel each other and certain kinds of materials.

All of the pages from Chapter 2 of the *Science on Target for Grade 3 Student Workbook*, are reproduced in this Parent/Teacher Edition in reduced-page format with sample answers. These activities will help your students develop the skills necessary to understand Physical Science.

Students should use the "Clues for Success" Checklist, for each activity in this section, as a tool to help them do their best work.

Activity 1

Step 1 Read the scenario "No School Today!"

No School Today!

The change in the winter weather came quickly! An ice storm hit the town!

The ice storm caused many changes in the lives of the students. First, the outdoors, including the roads and trees, looked like it was covered in glass. Next, the class had the day off from school. Lastly, staying home had many surprises because the electrical wires outdoors were not working properly. Electricity supplied by the power company traveled through these wires and supplied energy for many household uses.

Many of the daily needs the students expected were no longer available. Their homes were cold, if the only heat was provided by electricity. Some students were lucky because they had wood burning stoves or gas heaters. Most families ate food that was not heated. Watching TV was not possible. Batteries could be used for laptop computers, so students could email, watch DVDs, and listen to CDs for a while.

When night came, candles and flashlights were used for light. So, families played board games while they had enough light to see. Then, in the dark, they told stories and jokes!

Note: Student answers may vary. Example responses are for use as a guide.

Chapter 2

Physical Science

The activities in this section of the book will focus on Physical Science.

These activities will help you investigate ideas about objects and materials that include:

- observing and describing their physical properties,
- describing their position and changes of position, and
- examining ideas about light, heat, electricity, and magnetism.

Use the "Clues for Success" Checklist, as you complete each activity in this section, as a tool to help you do your best work.

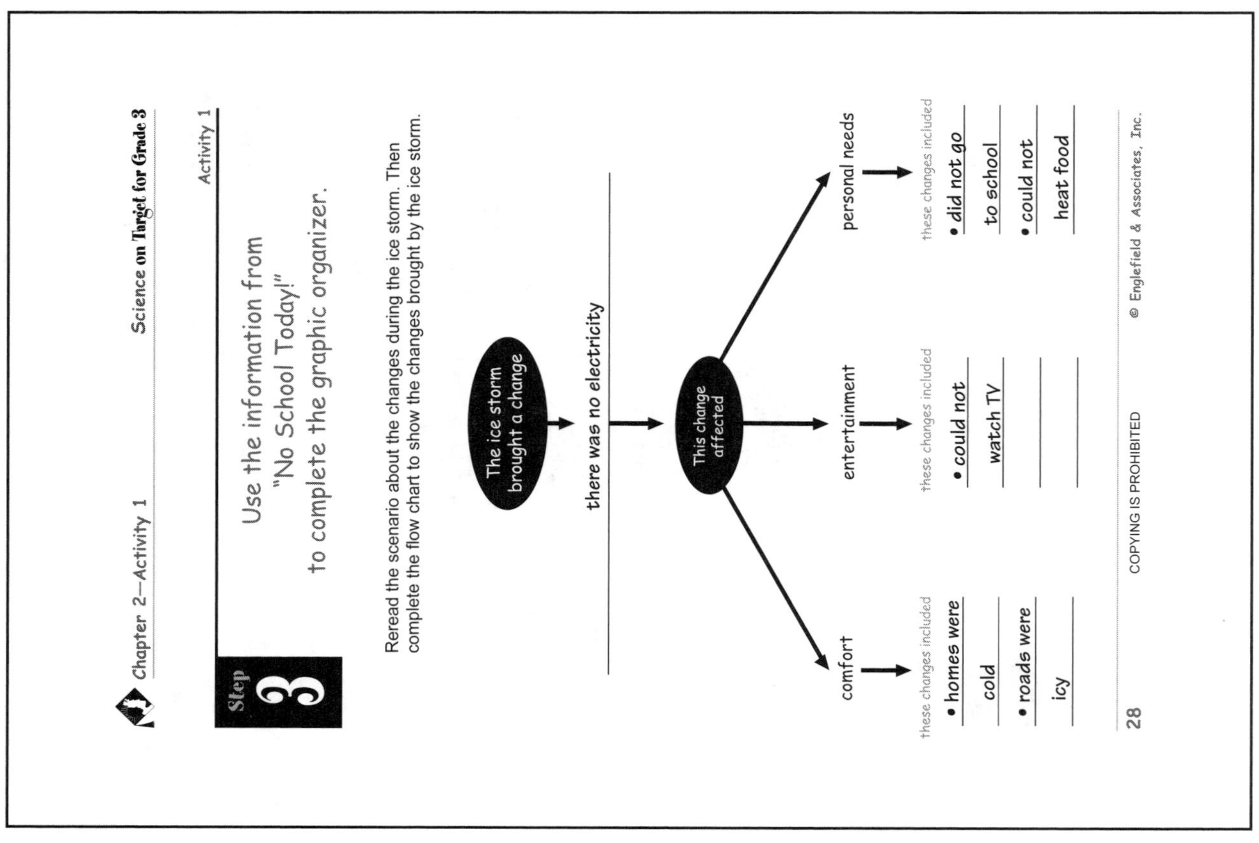

Step 3

Use the information from "No School Today!" to complete the graphic organizer.

Reread the scenario about the changes during the ice storm. Then complete the flow chart to show the changes brought by the ice storm.

Note: Student answers may vary. Example responses are for use as a guide.

Step 2

Complete the Checklist "Clues for Success."

The checklist will help you to read and think like a scientist.

Clues for Success

- ☐ **C**arefully read the information.
- ☐ **L**ook at any illustrations or diagrams.
  They may provide you with additional information to answer the question.
- ☐ **U**nderstand the way you are asked to answer the question.
  - ☐ Graph
  - ☐ Chart
  - ☐ Diagram
  - ☐ Complete sentences
  - ☐ Phrase
  - ☐ Filled circle
- ☐ **E**xamine the information given.
  - ☐ Reread the questions.
  - ☐ Underline key words or phrases.
  - ☐ Think about what the questions are asking.
- ☐ **S**ee if your answers match the questions.

Step 1

Activity 2

Read the scenario "Magnets."

Magnets

In art class the students were constructing games. Cleo created a maze with a cardboard box, paper, and markers. Her directions were attached to the maze.

Note: Student answers may vary. Example responses are for use as a guide.

Step 4

Activity 1

Answer the following questions for "No School Today!" using information from your graphic organizer.

If you were in an ice storm, what might you do in each situation listed?

1. There is no heat in your home *I would put on layers of clothes.*

2. There are no lights at night *I would use a flashlight or candles to help me see.*

3. You could not heat food *I would eat food that is good to eat at room temperature.*

4. You could not watch TV *I would read, play games, or clean my room.*

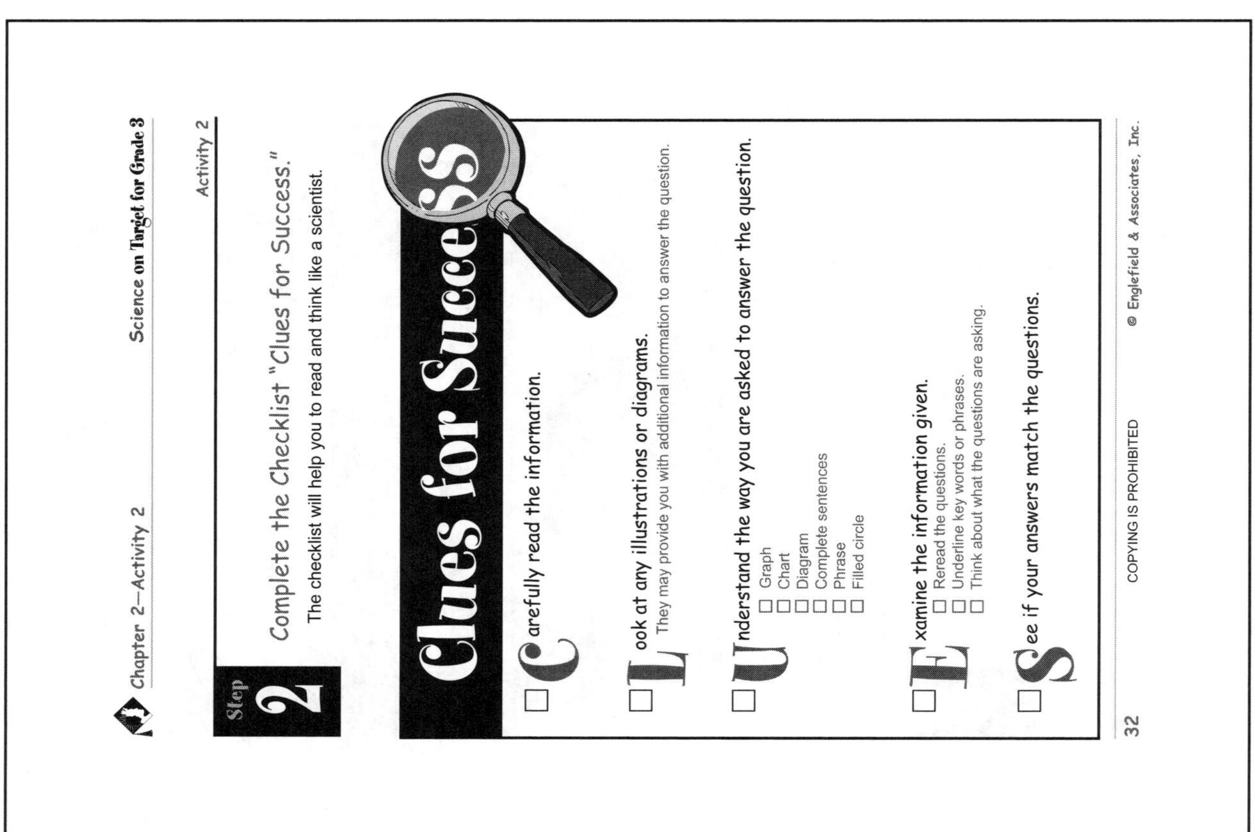

Step 2 Complete the Checklist "Clues for Success."   Activity 2

The checklist will help you to read and think like a scientist.

Clues for Success

- ☐ **C**arefully read the information.
- ☐ **L**ook at any illustrations or diagrams.
  They may provide you with additional information to answer the question.
- ☐ **U**nderstand the way you are asked to answer the question.
  - ☐ Graph
  - ☐ Chart
  - ☐ Diagram
  - ☐ Complete sentences
  - ☐ Phrase
  - ☐ Filled circle
- ☐ **E**xamine the information given.
  - ☐ Reread the questions.
  - ☐ Underline key words or phrases.
  - ☐ Think about what the questions are asking.
- ☐ **S**ee if your answers match the questions.

Note: Student answers may vary. Example responses are for use as a guide.

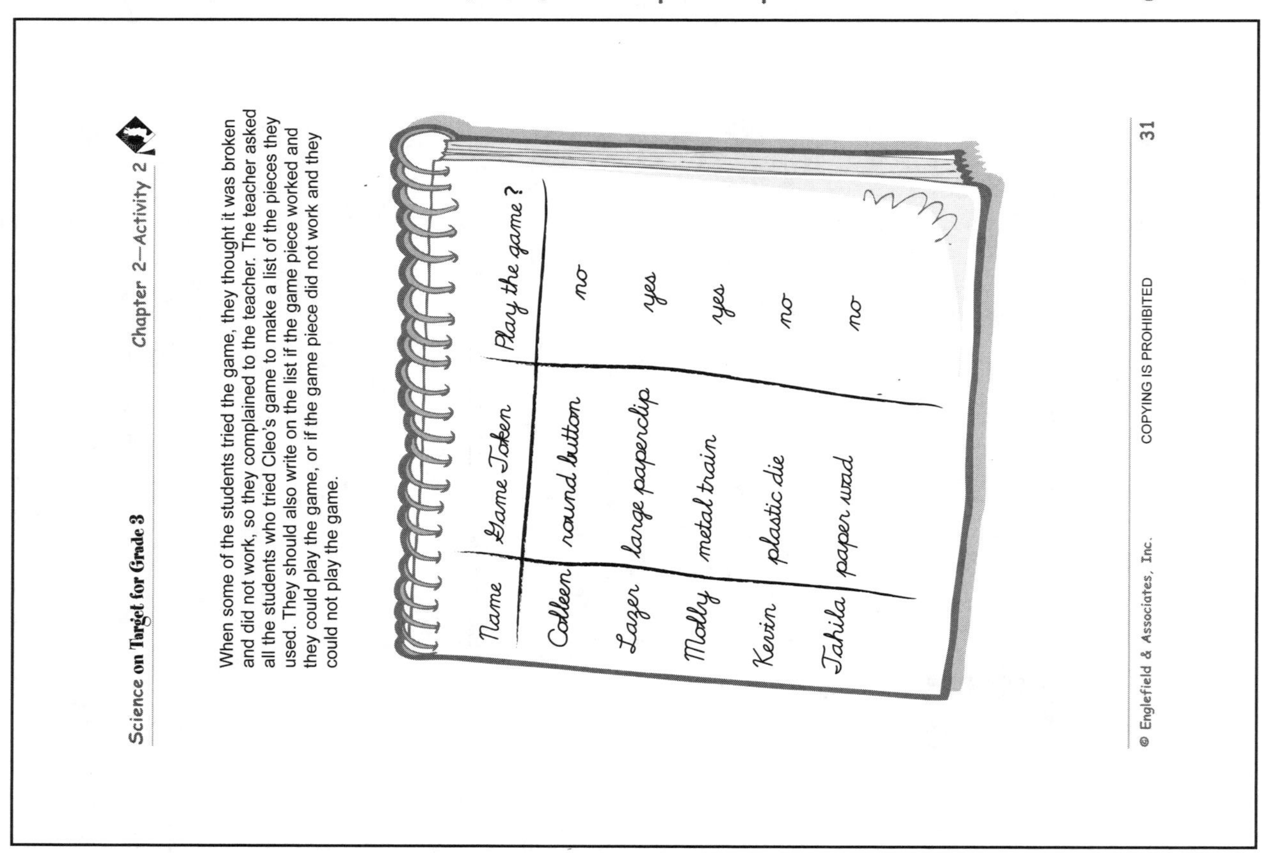

When some of the students tried the game, they thought it was broken and did not work, so they complained to the teacher. The teacher asked all the students who tried Cleo's game to make a list of the pieces they used. They should also write on the list if the game piece worked and they could play the game, or if the game piece did not work and they could not play the game.

Name	Game Token	Play the game?
Colleen	round button	no
Lazer	large paperclip	yes
Molly	metal train	yes
Kevin	plastic die	no
Tahila	paper wad	no

Step 4

Answer the following questions for "Magnets" using information from your graphic organizer.

Use the information on the chart to decide if the following game pieces could be used to play Cleo's game.

1. Small nail

● yes ○ no

Why?

It is made of metal so the magnet would move it.

2. Wad of clay

○ yes ● no

Why?

Clay is not metal. It would not be attracted to the magnet.

Note: Student answers may vary. Example responses are for use as a guide.

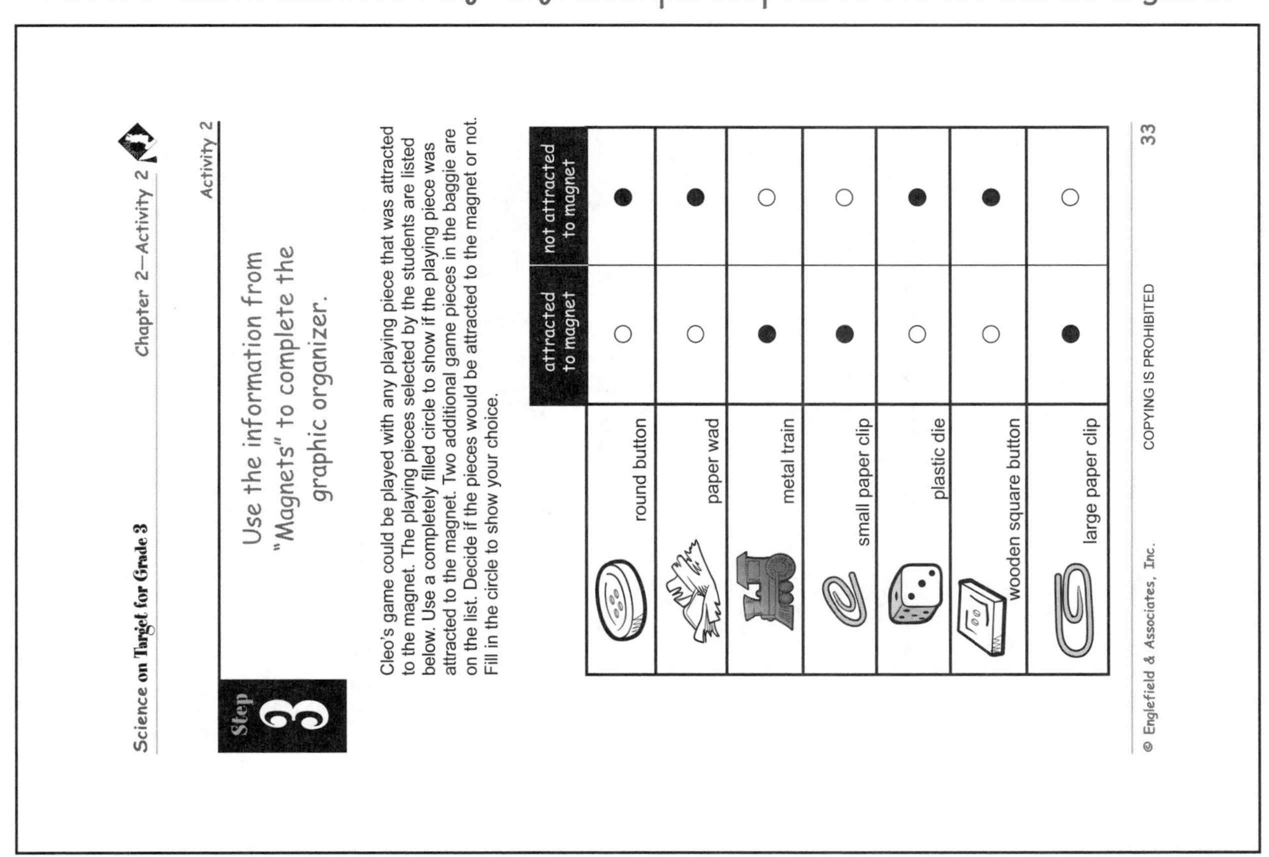

Step 3

Use the information from "Magnets" to complete the graphic organizer.

Cleo's game could be played with any playing piece that was attracted to the magnet. The playing pieces selected by the students are listed below. Use a completely filled circle to show if the playing piece was attracted to the magnet. Two additional game pieces in the baggie are on the list. Decide if the pieces would be attracted to the magnet or not. Fill in the circle to show your choice.

	attracted to magnet	not attracted to magnet
round button	○	●
paper wad	○	●
metal train	●	○
small paper clip	●	○
plastic die	○	●
wooden square button	○	●
large paper clip	●	○

Activity 3

Step 1

Read the scenario
"Heat It Up!"

Heat It Up!

The class was learning ways to keep warm for their winter campout. They discussed the different ways of keeping warm, like wearing warm clothes, layering clothing, and wearing hats and gloves. They also had a discussion of other ways to generate heat. **Heating** is the increase of energy in the small particles (atoms) that make up a substance. These particles are always in motion. The faster the particles move, the more thermal energy they have. The students also learned that this flow of thermal energy is called **conduction**. Conduction travels from the object with more energy (hotter) to the object with less energy (cooler).

Two ways of increasing the energy are rubbing and burning.

Rubbing is the act of creating energy (friction) by moving objects together or over each other, like moving your hands over each other to warm them. Hands feel warm because the action makes the particles in the hands move faster.

Burning is the reaction that gives off light and heat. It is also known as combustion. There are three conditions that are necessary for materials to burn. They are fuel, heat, and oxygen.

1. **Fuel** is the material that can burn.
2. **Heat** is the source that raises the temperature of the material to the point of burning.
3. There must be enough **oxygen** in the air to keep the reaction going.

The reaction continues until one of the needs is used up.

The students used these science ideas to keep warm during their campout!

Note: Student answers may vary. Example responses are for use as a guide.

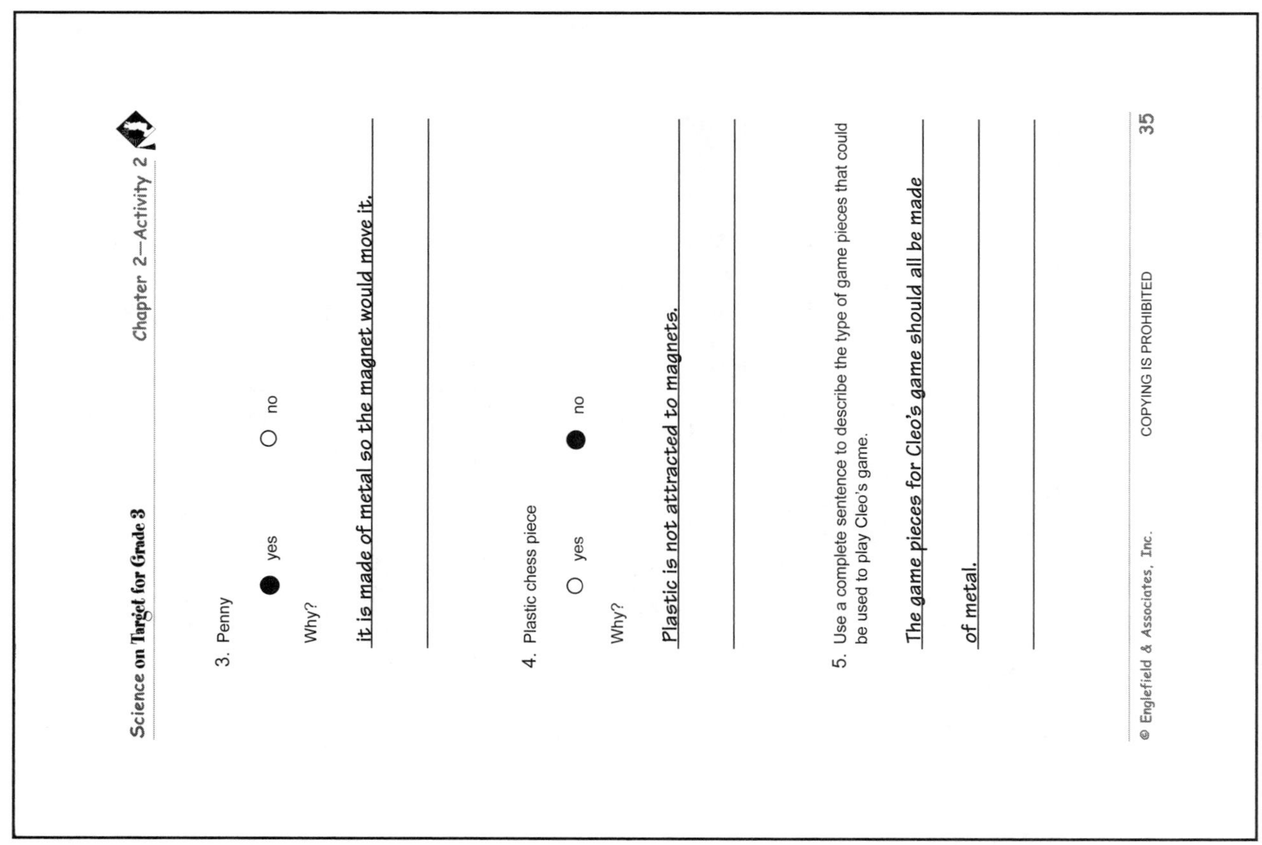

3. Penny

● yes ○ no

Why?

it is made of metal so the magnet would move it.

4. Plastic chess piece

○ yes ● no

Why?

Plastic is not attracted to magnets.

5. Use a complete sentence to describe the type of game pieces that could be used to play Cleo's game.

The game pieces for Cleo's game should all be made of metal.

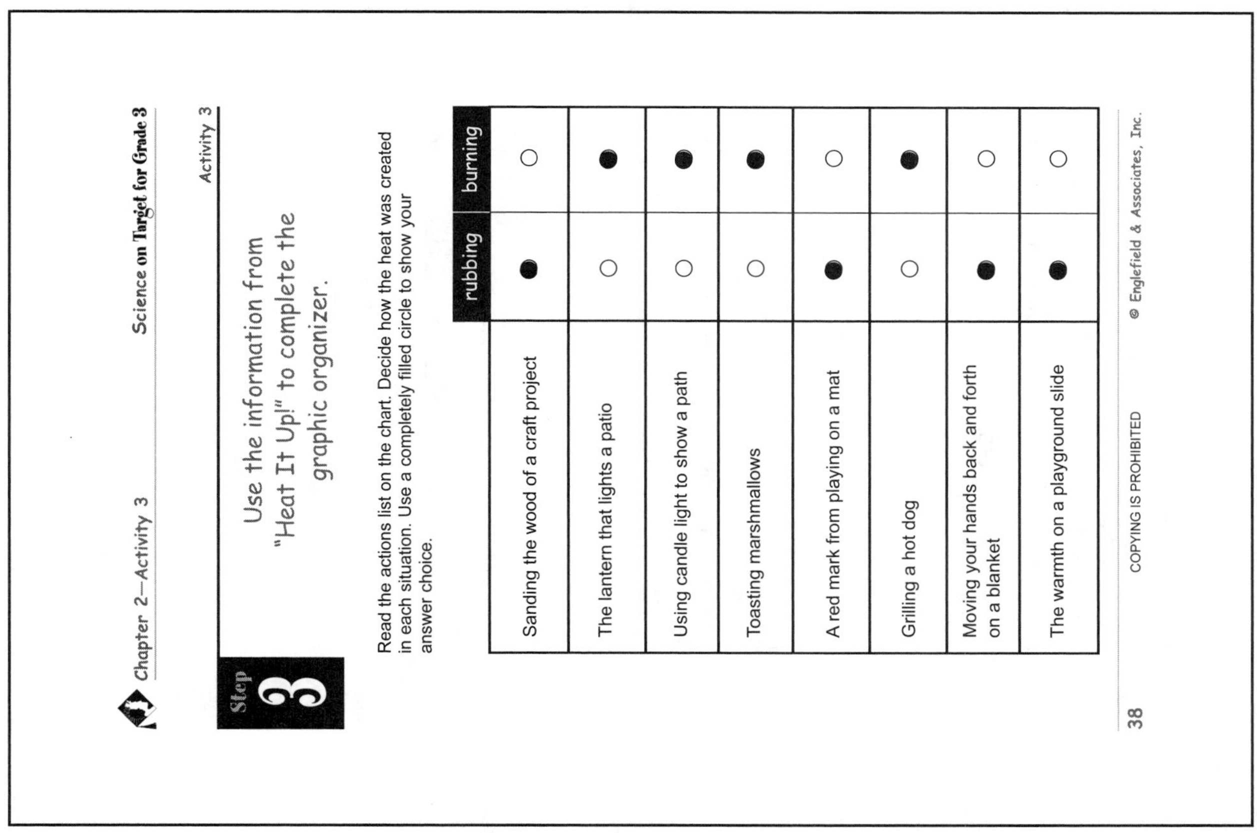

Activity 3

Step 3 Use the information from "Heat It Up!" to complete the graphic organizer.

Read the actions list on the chart. Decide how the heat was created in each situation. Use a completely filled circle to show your answer choice.

	rubbing	burning
Sanding the wood of a craft project	●	○
The lantern that lights a patio	○	●
Using candle light to show a path	○	●
Toasting marshmallows	○	●
A red mark from playing on a mat	●	○
Grilling a hot dog	○	●
Moving your hands back and forth on a blanket	●	○
The warmth on a playground slide	●	○

Note: Student answers may vary. Example responses are for use as a guide.

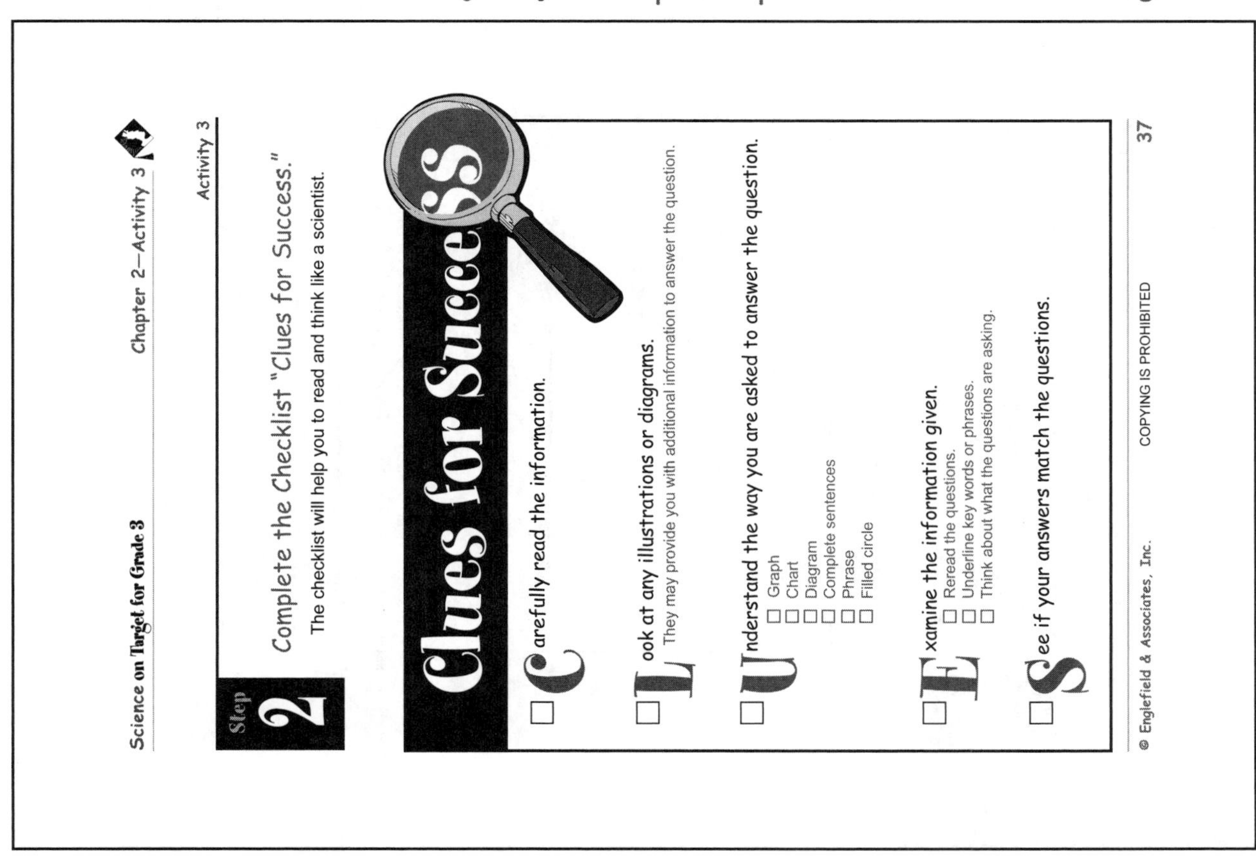

Activity 3

Step 2 Complete the Checklist "Clues for Success."

The checklist will help you to read and think like a scientist.

Clues for Success

- ☐ **C**arefully read the information.
- ☐ **L**ook at any illustrations or diagrams.
  They may provide you with additional information to answer the question.
- ☐ **U**nderstand the way you are asked to answer the question.
  - ☐ Graph
  - ☐ Chart
  - ☐ Diagram
  - ☐ Complete sentences
  - ☐ Phrase
  - ☐ Filled circle
- ☐ **E**xamine the information given.
  - ☐ Reread the questions.
  - ☐ Underline key words or phrases.
  - ☐ Think about what the questions are asking.
- ☐ **S**ee if your answers match the questions.

Step 4 Activity 3

Answer the following questions for "Heat It Up!" using information from your graphic organizer.

1. In the space provided, draw a triangle. Then, write a need for a fire (burning) on each corner.

Fuel (wood, etc.)

Oxygen (air)

Heat (match, spark)

Each team of students on the camp out had difficulty building their fire. Read the description of each team's fire pit and write a complete sentence to tell which condition for making a fire was not met.

2. Team 1: The team packed the logs and twigs airtight.

   There was not enough oxygen for the fire.

3. Team 2: The team used the largest logs they could find to build their fire.

   They needed smaller fuel to start the fire.

4. Team 3: The team rubbed twigs together to start their fire.

   The rubbing did not create enough energy/heat to start the fire.

5. Team 4: The team built their fire with finger-sized twigs.

   The twigs were not enough fuel to keep the fire going once it got started.

Note: Student answers may vary. Example responses are for use as a guide.

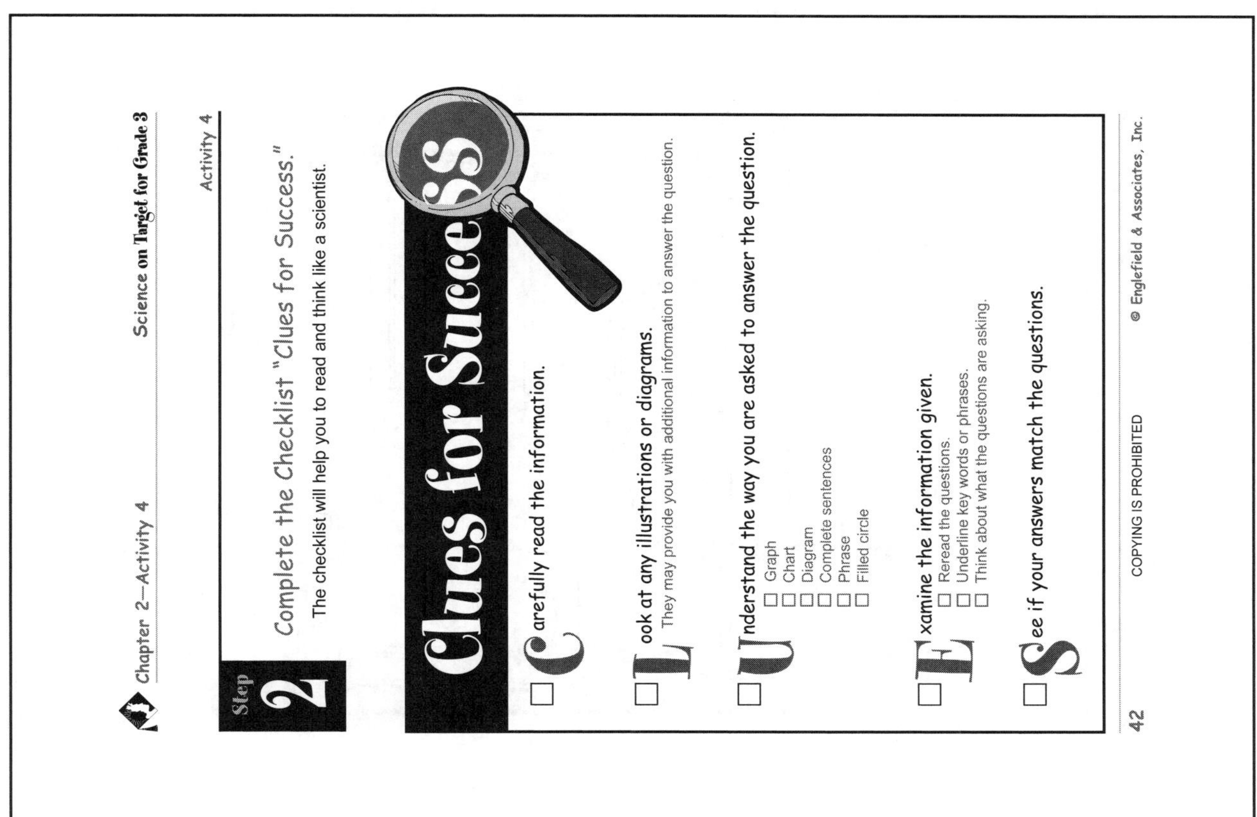

Chapter 2—Activity 4 Science on Target for Grade 3

Step 2

Complete the Checklist "Clues for Success."

The checklist will help you to read and think like a scientist.

Activity 4

Clues for Success

☐ **C**arefully read the information.

☐ **L**ook at any illustrations or diagrams.
They may provide you with additional information to answer the question.

☐ **U**nderstand the way you are asked to answer the question.
- ☐ Graph
- ☐ Chart
- ☐ Diagram
- ☐ Complete sentences
- ☐ Phrase
- ☐ Filled circle

☐ **E**xamine the information given.
- ☐ Reread the questions.
- ☐ Underline key words or phrases.
- ☐ Think about what the questions are asking.

☐ **S**ee if your answers match the questions.

42 COPYING IS PROHIBITED © Englefield & Associates, Inc.

Note: Student answers may vary. Example responses are for use as a guide.

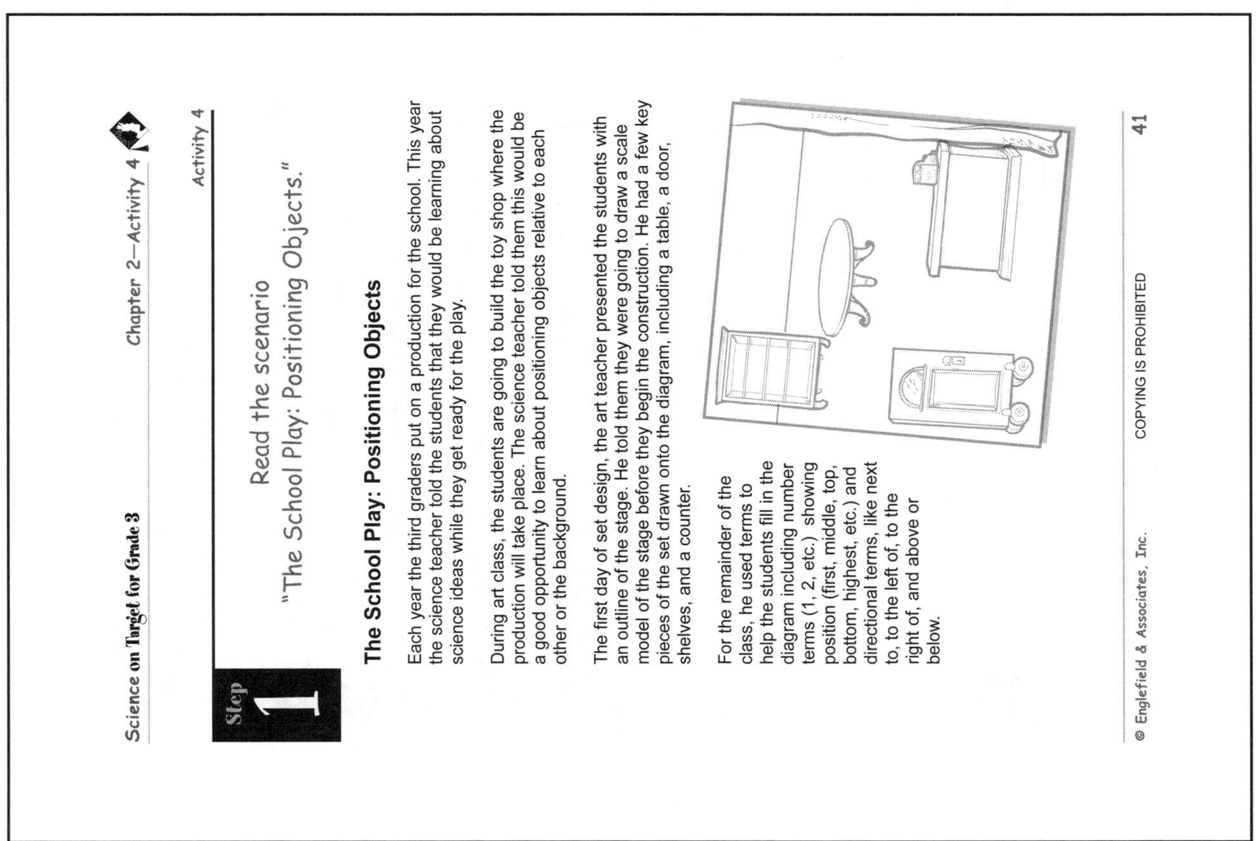

Science on Target for Grade 3 Chapter 2—Activity 4

Step 1

Read the scenario "The School Play: Positioning Objects."

Activity 4

The School Play: Positioning Objects

Each year the third graders put on a production for the school. This year the science teacher told the students that they would be learning about science ideas while they get ready for the play.

During art class, the students are going to build the toy shop where the production will take place. The science teacher told them this would be a good opportunity to learn about positioning objects relative to each other or the background.

The first day of set design, the art teacher presented the students with an outline of the stage. He told them they were going to draw a scale model of the stage before they begin the construction. He had a few key pieces of the set drawn onto the diagram, including a table, a door, shelves, and a counter.

For the remainder of the class, he used terms to help the students fill in the diagram including number terms (1, 2, etc.) showing position (first, middle, top, bottom, highest, etc.) and directional terms, like next to, to the left of, to the right of, and above or below.

© Englefield & Associates, Inc. COPYING IS PROHIBITED 41

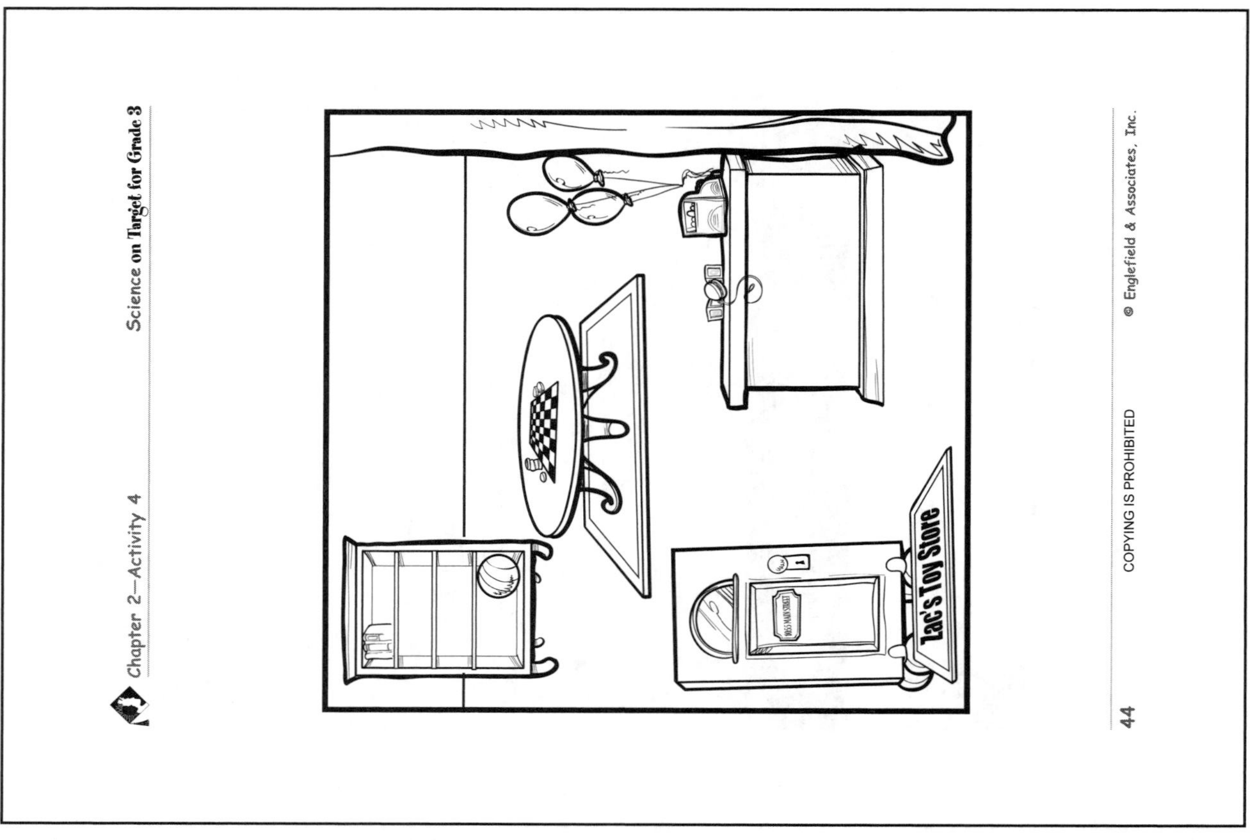

Chapter 2—Activity 4

Science on Target for Grade 3

44 COPYING IS PROHIBITED © Englefield & Associates, Inc.

Note: Student answers may vary. Example responses are for use as a guide.

Science on Target for Grade 3

Chapter 2—Activity 4

Activity 4

Step 3

Use the information from "The School Play: Positioning Objects" to complete the graphic organizer.

To complete a diagram of the set for the play, draw the objects listed below on the scale model of the stage on page 44. Be sure to draw your objects in the position given.

Objects to Add to the Stage Model:

1. Draw 3 books on the highest shelf.
2. Place a square rug under the table.
3. Put a sign on the middle of the door with the address 1055 Main Street.
4. Draw a rectangular mat outside the door with the store's name: "Zac's Toy Store".
5. Add 2 blocks with a yoyo between them on the counter.
6. In the corner behind the counter, draw 3 round helium filled balloons.
7. Draw a checkers game on the table.
8. Add a ball to the bottom shelf.

© Englefield & Associates, Inc. COPYING IS PROHIBITED 43

Activity 4

Step 4

Answer the following questions for "The School Play: Positioning Objects" using information from your graphic organizer.

In the second scene of the play, the toy shop owner is going to add more items to the toy store. Read the type of item, then make a suggestion to tell where the owner might place the item. Then, use a complete sentence to state why it should be placed there.

1. **Item: 5 new books**
   Where do you think they should be placed?

   On the book shelf.

   Why?

   They should be placed in the area near the other books.

2. **Item: 2 wagons**
   Where do you think they should be placed?

   On the back wall behind the table.

   Why?

   There is enough room for both wagons along the back wall.

Note: Student answers may vary. Example responses are for use as a guide.

3. **Item: A stack of papers advertising a special program at the library**
   Where do you think they should be placed?

   On the counter by the cash register.

   Why?

   They should be near the cash register so people can pick one up when they pay for their purchase.

4. **Item: A poster about a bicycle safety class at the fire station**
   Where do you think it should be placed?

   On the wall behind the table.

   Why?

   The poster should be placed on the wall so everyone can see it easily.

Step 2

Complete the Checklist "Clues for Success."

The checklist will help you to read and think like a scientist.

Clues for Success

- ☐ **C**arefully read the information.
- ☐ **L**ook at any illustrations or diagrams.
  They may provide you with additional information to answer the question.
- ☐ **U**nderstand the way you are asked to answer the question.
  - ☐ Graph
  - ☐ Chart
  - ☐ Diagram
  - ☐ Complete sentences
  - ☐ Phrase
  - ☐ Filled circle
- ☐ **E**xamine the information given.
  - ☐ Reread the questions.
  - ☐ Underline key words or phrases.
  - ☐ Think about what the questions are asking.
- ☐ **S**ee if your answers match the questions.

Note: Student answers may vary. Example responses are for use as a guide.

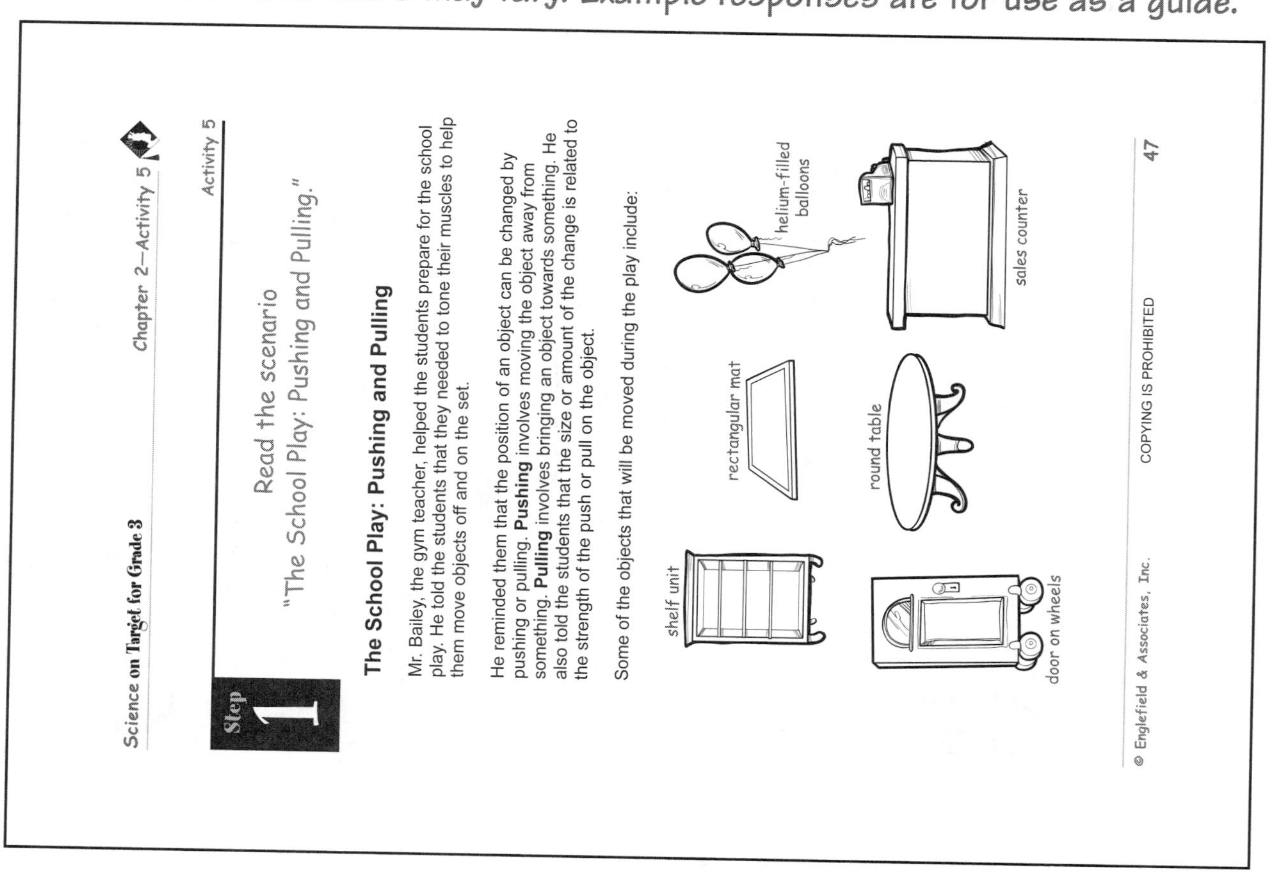

Step 1

Read the scenario
"The School Play: Pushing and Pulling."

The School Play: Pushing and Pulling

Mr. Bailey, the gym teacher, helped the students prepare for the school play. He told the students that they needed to tone their muscles to help them move objects off and on the set.

He reminded them that the position of an object can be changed by pushing or pulling. **Pushing** involves moving the object away from something. **Pulling** involves bringing an object towards something. He also told the students that the size or amount of the change is related to the strength of the push or pull on the object.

Some of the objects that will be moved during the play include:

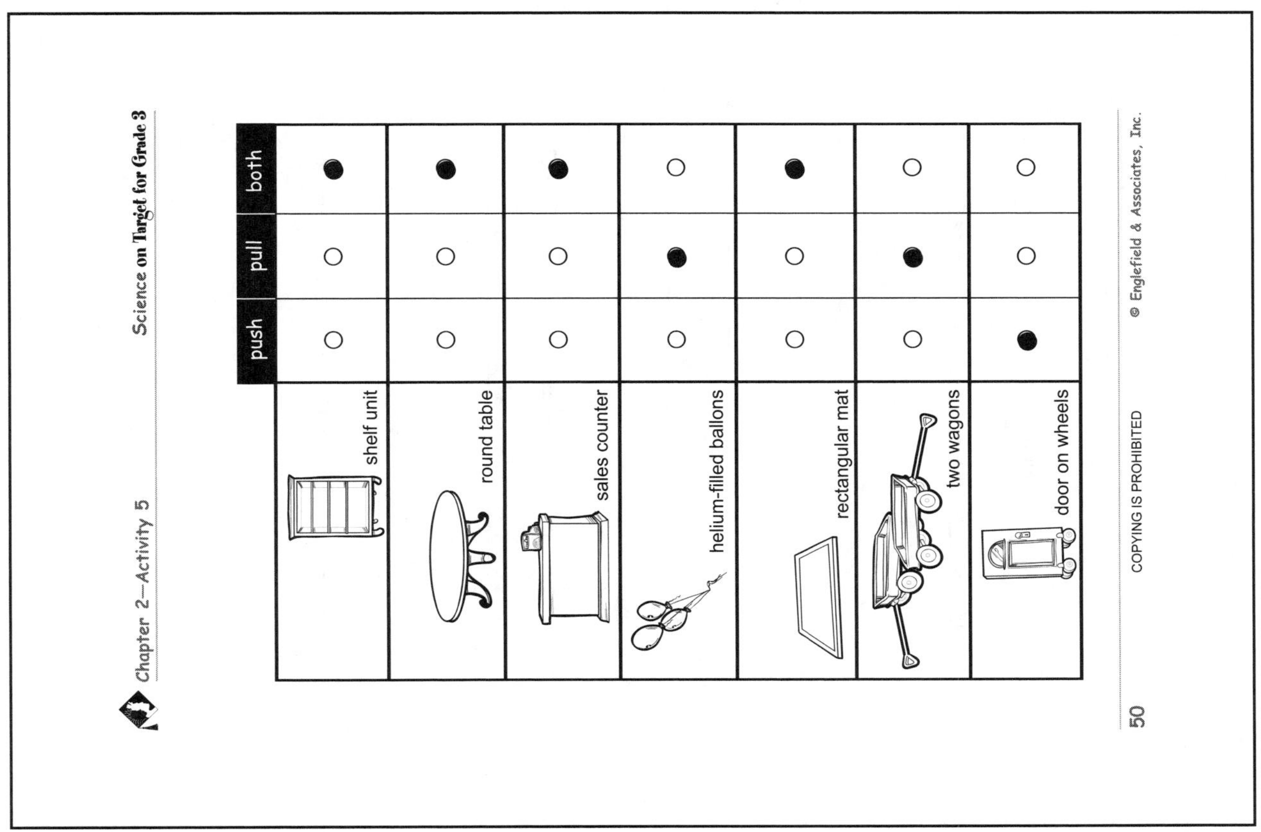

	push	pull	both
shelf unit	○	○	●
round table	○	○	●
sales counter	○	○	●
helium-filled ballons	○	●	○
rectangular mat	○	○	●
two wagons	○	●	○
door on wheels	●	○	○

Note: Student answers may vary. Example responses are for use as a guide.

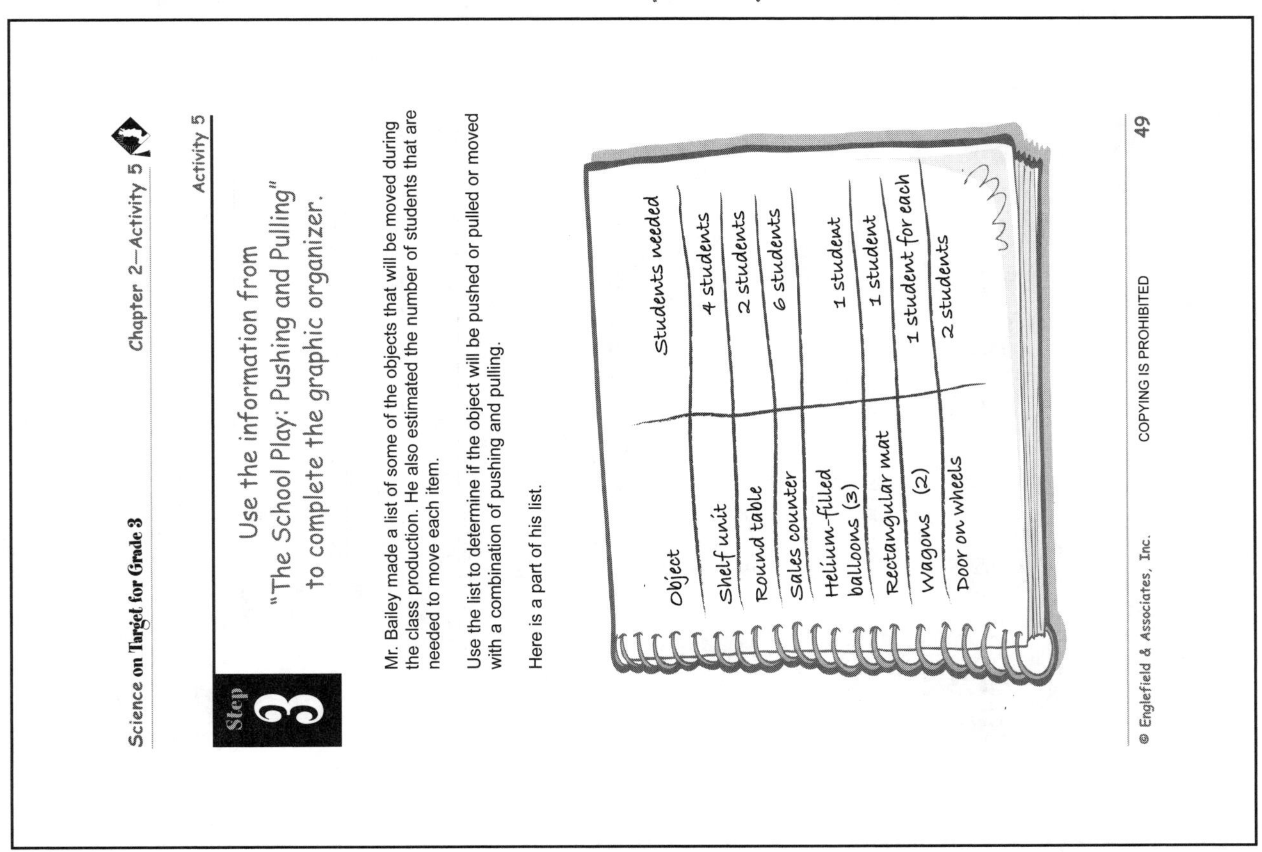

Step 3

Activity 5

Use the information from "The School Play: Pushing and Pulling" to complete the graphic organizer.

Mr. Bailey made a list of some of the objects that will be moved during the class production. He also estimated the number of students that are needed to move each item.

Use the list to determine if the object will be pushed or pulled or moved with a combination of pushing and pulling.

Here is a part of his list.

Object	Students needed
Shelf unit	4 students
Round table	2 students
Sales counter	6 students
Helium-filled balloons (3)	1 student
Rectangular mat	1 student
Wagons (2)	1 student for each
Door on wheels	2 students

2. The door for the scene is on wheels. Use a complete sentence to give a reason that the door is on wheels.

The door is very heavy. It is on wheels to make it easier to move.

3. The balloons are filled with helium. How might the students make sure they don't float off the stage?

The students should add weights to the balloons' strings or tie them onto something so they don't float away.

Note: Student answers may vary. Example responses are for use as a guide.

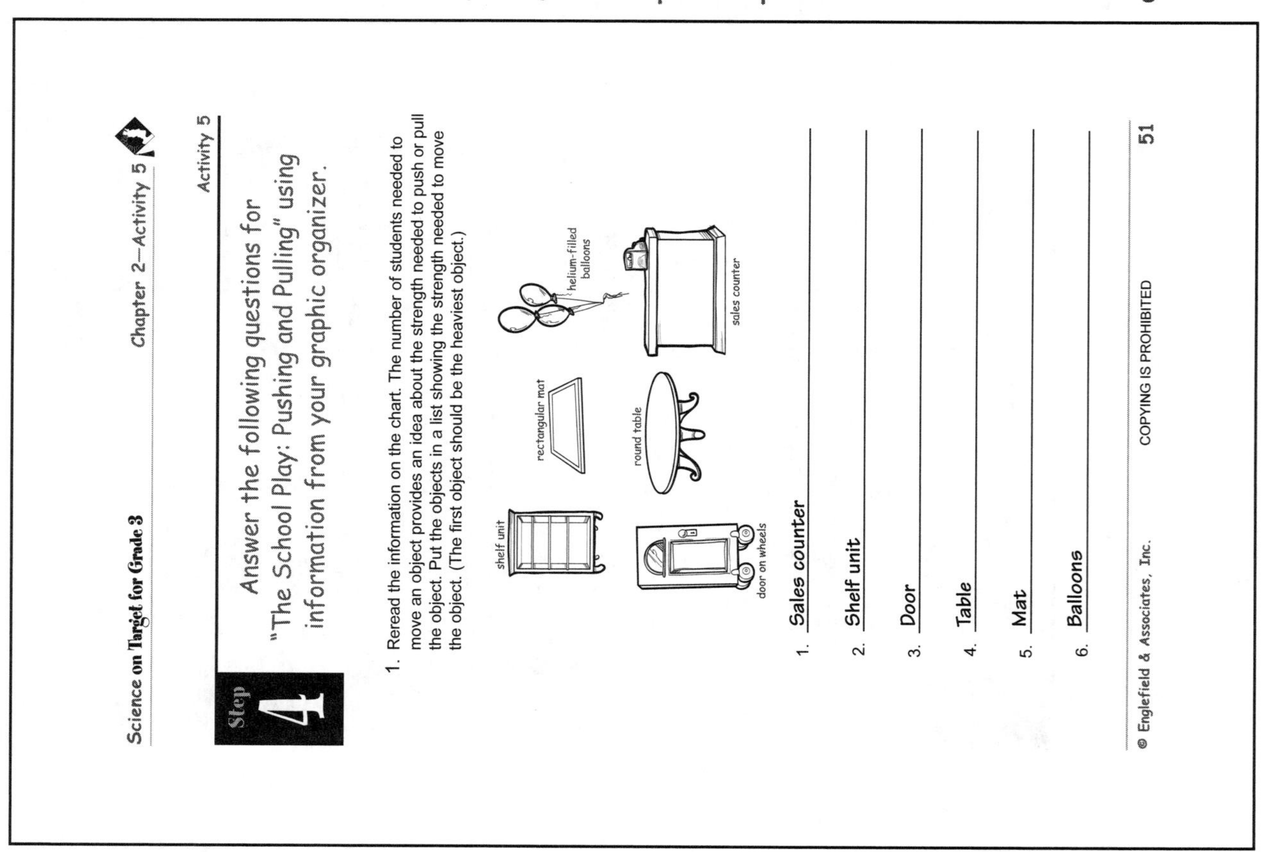

Activity 5

Step 4 Answer the following questions for "The School Play: Pushing and Pulling" using information from your graphic organizer.

1. Reread the information on the chart. The number of students needed to move an object provides an idea about the strength needed to push or pull the object. Put the objects in a list showing the strength needed to move the object. (The first object should be the heaviest object.)

1. *Sales counter*
2. *Shelf unit*
3. *Door*
4. *Table*
5. *Mat*
6. *Balloons*

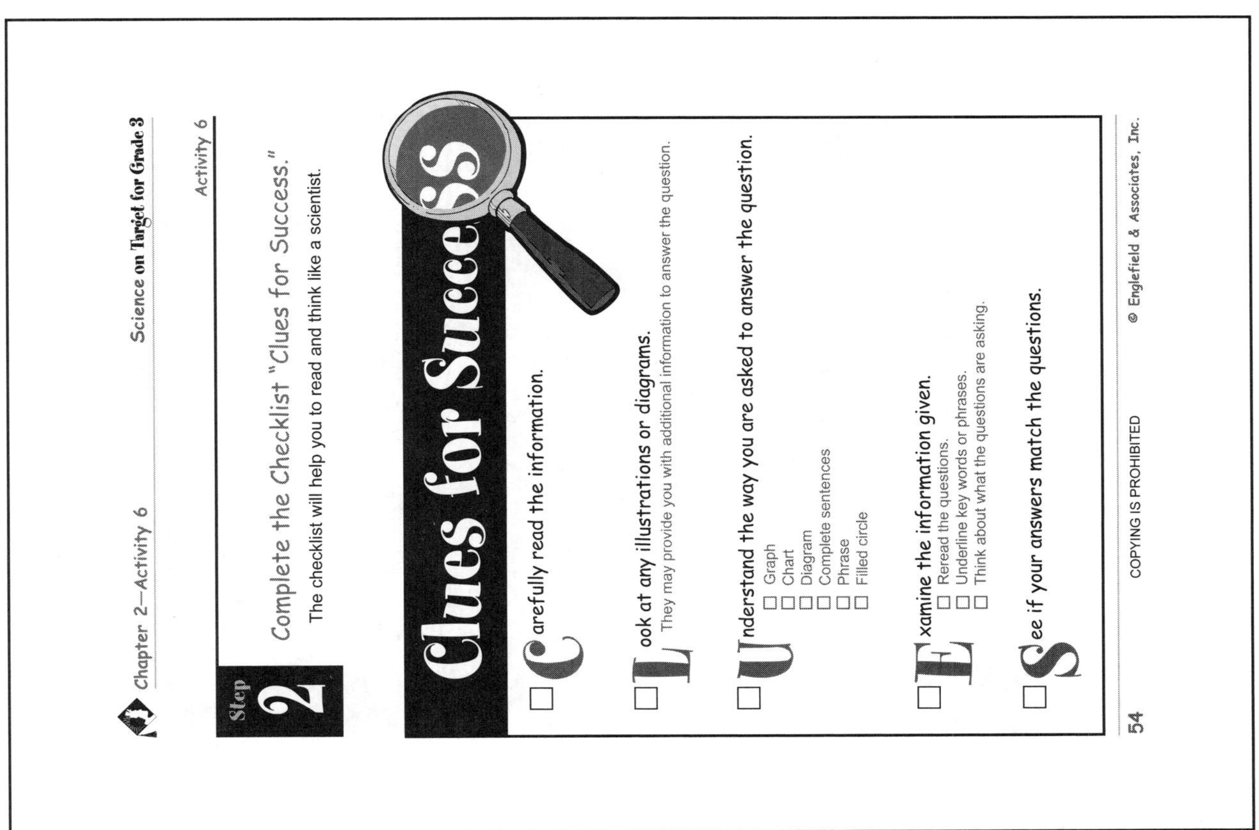
Science on Target for Grade 3

Chapter 2—Activity 6

Activity 6

Step 2

Complete the Checklist "Clues for Success."

The checklist will help you to read and think like a scientist.

Clues for Success

☐ Carefully read the information.

☐ Look at any illustrations or diagrams.
They may provide you with additional information to answer the question.

☐ Understand the way you are asked to answer the question.
☐ Graph
☐ Chart
☐ Diagram
☐ Complete sentences
☐ Phrase
☐ Filled circle

☐ Examine the information given.
☐ Reread the questions.
☐ Underline key words or phrases.
☐ Think about what the questions are asking.

☐ See if your answers match the questions.

54 COPYING IS PROHIBITED © Englefield & Associates, Inc.

Note: Student answers may vary. Example responses are for use as a guide.

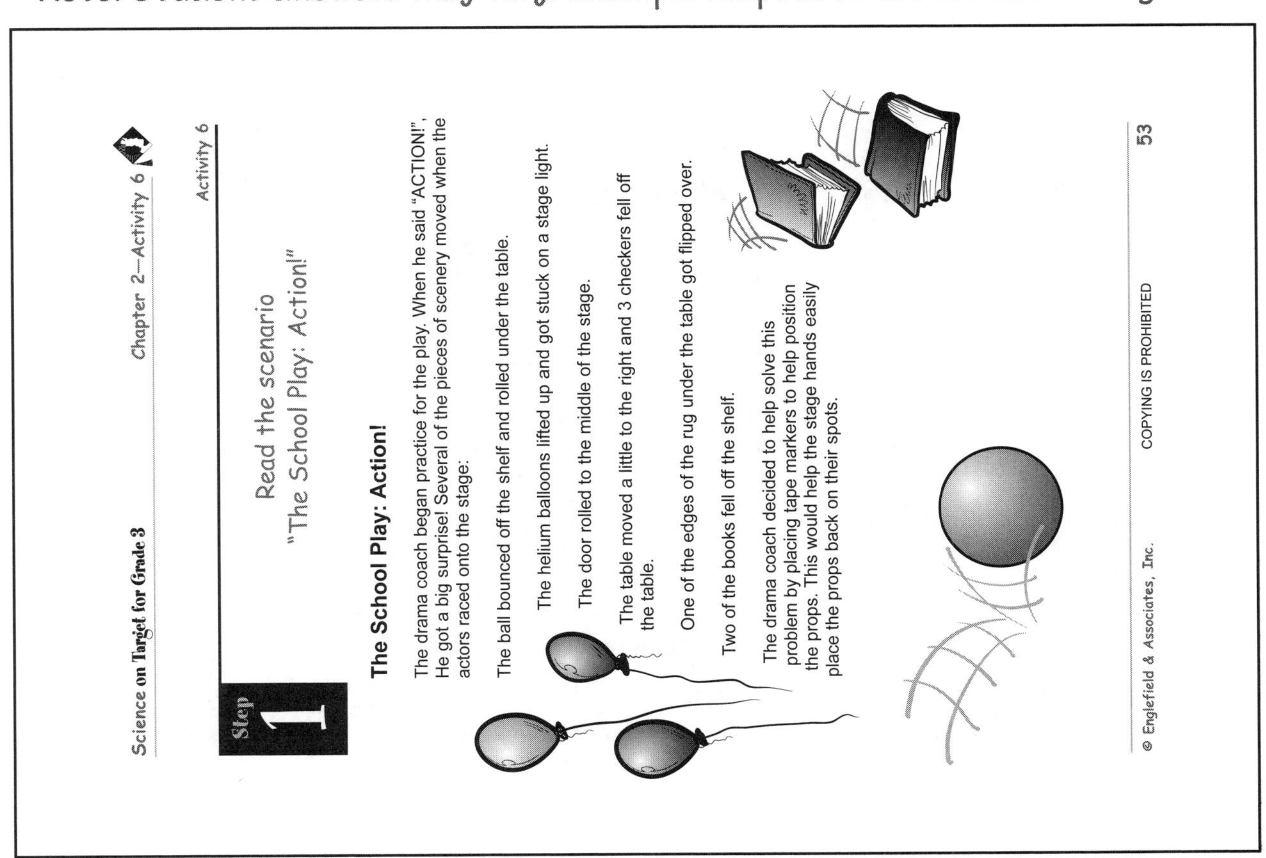
Science on Target for Grade 3

Chapter 2—Activity 6

Activity 6

Step 1

Read the scenario "The School Play: Action!"

The School Play: Action!

The drama coach began practice for the play. When he said "ACTION!", He got a big surprise! Several of the pieces of scenery moved when the actors raced onto the stage:

The ball bounced off the shelf and rolled under the table.

The helium balloons lifted up and got stuck on a stage light.

The door rolled to the middle of the stage.

The table moved a little to the right and 3 checkers fell off the table.

One of the edges of the rug under the table got flipped over.

Two of the books fell off the shelf.

The drama coach decided to help solve this problem by placing tape markers to help position the props. This would help the stage hands easily place the props back on their spots.

© Englefield & Associates, Inc. COPYING IS PROHIBITED 53

Activity 6

Step 3

Use the information from "The School Play: Action!" to complete the graphic organizer.

Reread the scenario to help you add these objects to their place on the set before "Action" is called on the stage. Cross out each object below when you draw it.

ball balloons checker game mat books

Note: Student answers may vary. Example responses are for use as a guide.

Activity 6

Step 4

Answer the following questions for "The School Play: Action!" using information from your graphic organizer.

1. Reread the changes to the set that happened when the director said "ACTION!" On page 55, put a small dot next to the object, then use arrows to trace the path and direction that each object moved. Use a line to show the path and arrows to show the direction the object moved.

2. Some objects moved when action began on the set of the play. Other objects stayed in their place. Use a complete sentence to answer the questions below.

   Why did the ball move?

   The ball moved because it is round and the action on the stage created a change in energy to help it move.

   What might have made the mat corner flip up?

   Someone might have kicked the corner of the mat.

   Why didn't the table move?

   The table didn't move because it is heavy.

Chapter 3

Chapter 3 of the *Science on Target for Grade 3 Student Workbook*, covers the National Content Standards for Life Science. The standards are as follows:

Life Science

The characteristics of organisms

- Organisms have basic needs for life (animals need air, water, food; plants need air, water, nutrients, and light). Organisms can survive only in environments that meet their needs. There are many different environments that support different types of living things.
- Each plant and animal has different structures that serve different functions in growth, survival, and reproduction. (Humans have different structures for walking, holding, seeing, and talking.)
- The behavior of individual organisms is influenced by internal cues (such as hunger) and external cues (such as changes in the environment). Humans have senses that help them detect internal and external cues.

Life cycles of organisms

- Plants and animals have life cycles that include being born, developing into adults, reproducing, and eventually dying. The details of this life cycle are different for different organisms.
- Plants and animals closely resemble their parents.
- Many characteristics of an organism are inherited from the parents of the organism, but other characteristics result from an individual's interactions with the environment. Inherited characteristics include color of flowers and number of limbs of an animal. Other features, such as the ability to ride a bicycle, are learned through interactions with the environment and cannot be passed on to the next generation.

Organisms and environments

- All animals depend on plants. Some animals eat plants for food. Other animals eat animals that eat the plants.
- An organism's patterns of behavior are related to the nature of that organism's environment, including the kinds and numbers of other organisms present, the availability of food and resources, and the physical characteristics of the environment. When the environment changes, some plants and animals survive and reproduce, while others die or move on to new locations.
- All organisms cause changes in the environment where they live. Some of these changes are detrimental to the organisms, whereas others are beneficial.
- Humans depend on their natural and constructed environments. Humans change environments in ways that can be either beneficial or detrimental for themselves and other organisms.

All of the pages from Chapter 3 of the *Science on Target for Grade 3 Student Workbook*, are reproduced in this Parent/Teacher Edition in reduced-page format with sample answers. These activities will help your students develop the skills necessary to understand Life Science.

Students should use the "Clues for Success" Checklist, for each activity in this section, as a tool to help them do their best work.

Step 1

Activity 1

Read the scenario
"Characteristics of Individuals."

Characteristics of Individuals

In social studies, the class is working on a family tree assignment. Each student brought in photographs of their family members to share. Luke has photographs of his grandma, his uncle (his mother's brother), his mom and also his school picture. When he placed the photographs on the paper, he noticed that they looked very much alike.

These members of Luke's family shared many inherited characteristics, like dark curly hair, round-shaped faces, and the same wide grins.

Inherited characteristics are passed on from parents to their offspring in all organisms. Some inherited characteristics in organisms may include the color of flowers, the shape of leaves on a tree, and the color of a dog's coat.

Note: Student answers may vary. Example responses are for use as a guide.

Chapter 3

Life Science

The activities in this section of the book will focus on Life Science.

These activities will help you investigate:

- the characteristics of organisms and the way they meet their needs for life,
- the life cycles of organisms, and
- the ways organisms interact with their environment.

Use the "Clues for Success" Checklist, as you complete each activity in this section, as a tool to help you do your best work.

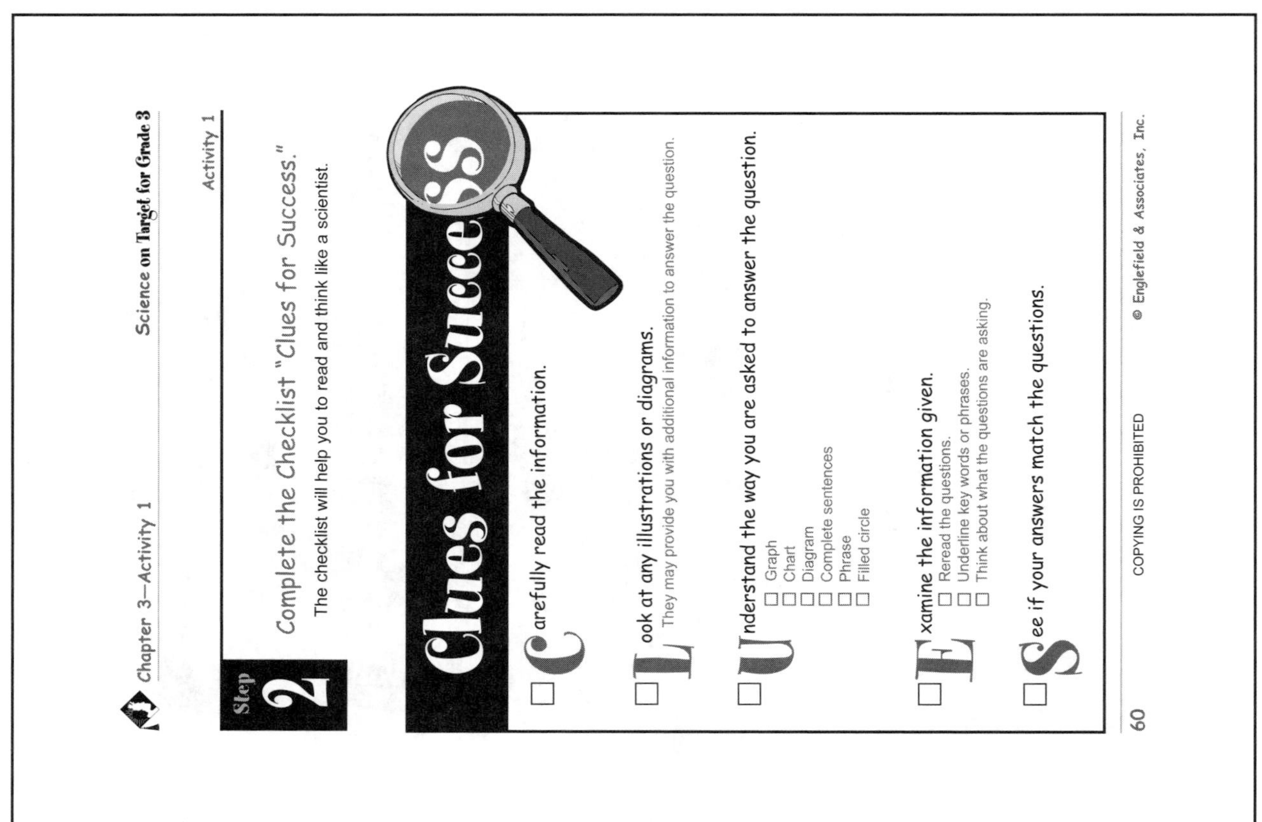

Chapter 3—Activity 1 Science on Target for Grade 3

Activity 1

Step 2 Complete the Checklist "Clues for Success."

The checklist will help you to read and think like a scientist.

Clues for Success

- ☐ **C**arefully read the information.
- ☐ **L**ook at any illustrations or diagrams.
  They may provide you with additional information to answer the question.
- ☐ **U**nderstand the way you are asked to answer the question.
  - ☐ Graph
  - ☐ Chart
  - ☐ Diagram
  - ☐ Complete sentences
  - ☐ Phrase
  - ☐ Filled circle
- ☐ **E**xamine the information given.
  - ☐ Reread the questions.
  - ☐ Underline key words or phrases.
  - ☐ Think about what the questions are asking.
- ☐ **S**ee if your answers match the questions.

60 COPYING IS PROHIBITED © Englefield & Associates, Inc.

Note: Student answers may vary. Example responses are for use as a guide.

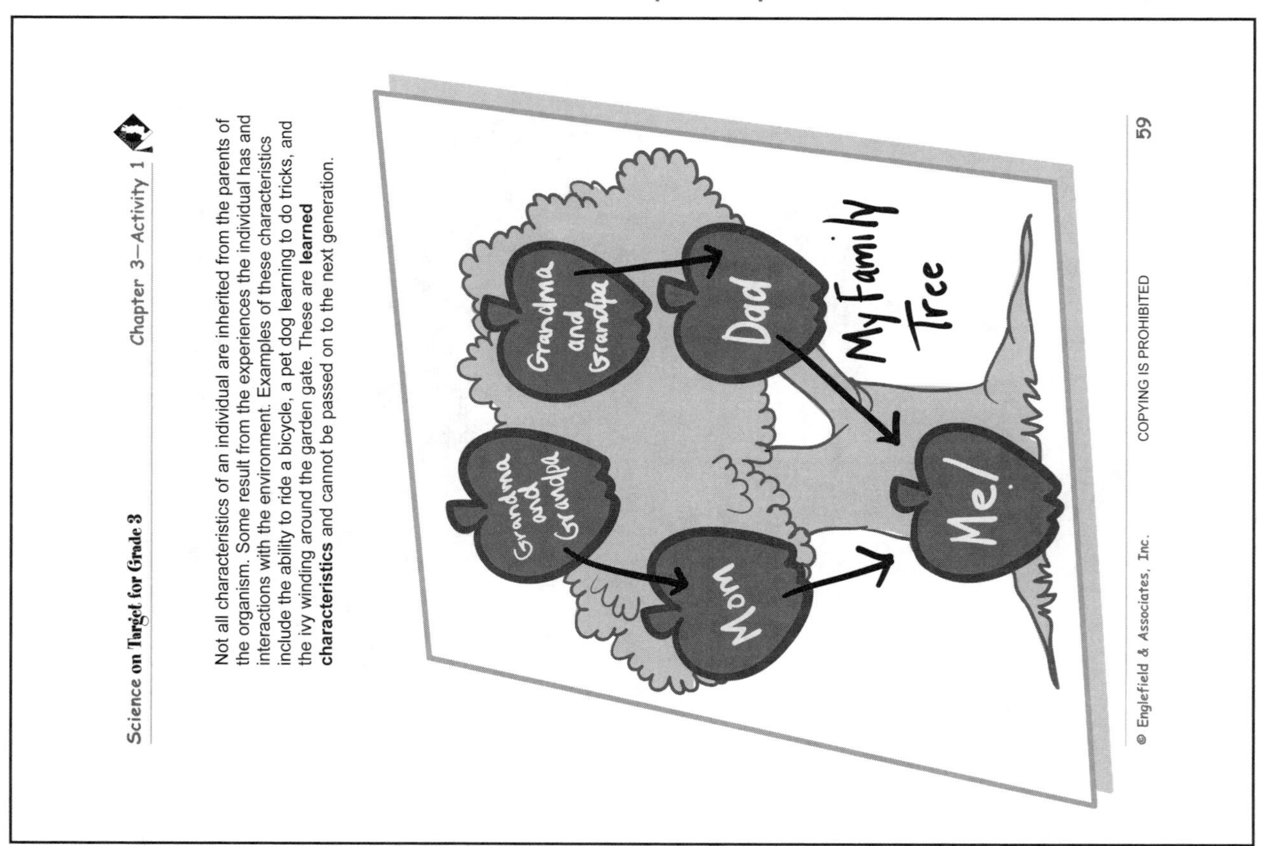

Science on Target for Grade 3 Chapter 3—Activity 1

Not all characteristics of an individual are inherited from the parents of the organism. Some result from the experiences the individual has and interactions with the environment. Examples of these characteristics include the ability to ride a bicycle, a pet dog learning to do tricks, and the ivy winding around the garden gate. These are **learned characteristics** and cannot be passed on to the next generation.

© Englefield & Associates, Inc. COPYING IS PROHIBITED 59

Note: Student answers may vary. Example responses are for use as a guide.

Activity 1

Step 3

Use the information from "Characteristics of Individuals" to complete the graphic organizer.

Reread the scenario, "Characteristics of Individuals." Then organize the information provided about the characteristics of an organism by completing the outline below.

Characteristics of Individuals

I. Inherited characteristics are passed from one generation to the next generation.

- A. Examples of this kind of characteristic found in Luke's family.
  - dark, curly hair
  - round-shaped face
  - grin
- B. Examples of this kind of characteristic found in
  1. Flowers color
  2. Dogs color of coat

II. Learned characteristics cannot be passed from one generation to the next generation.

They are the result of experiences or interactions with the environment.

- A. Examples of this kind of characteristic found in
  1. People riding a bicycle
  2. Plants winding around a garden gate
  3. Dogs learning tricks

Activity 1

Step 4

Answer the following questions for "Characteristics of Individuals" using information from your graphic organizer.

Luke made a list of his personal characteristics. Reread the information in the scenario and on the graphic organizer to help decide if the characteristics on his list are **inherited** or **learned**.

	inherited characteristic	learned characteristic
Luke has blue eyes like his grandfather.	●	○
Luke likes to read superhero comic books.	○	●
Luke has long fingers like his brothers and sister.	●	○
Luke practiced an hour every day to make the traveling ice hockey team.	○	●
Luke likes to read to the kindergarten class.	○	●

Chapter 3—Activity 2 Science on Target for Grade 3

Activity 2

Step 2

Complete the Checklist "Clues for Success."

The checklist will help you to read and think like a scientist.

Clues for Success

- ☐ **C**arefully read the information.
- ☐ **L**ook at any illustrations or diagrams.
  They may provide you with additional information to answer the question.
- ☐ **U**nderstand the way you are asked to answer the question.
  - ☐ Graph
  - ☐ Chart
  - ☐ Diagram
  - ☐ Complete sentences
  - ☐ Phrase
  - ☐ Filled circle
- ☐ **E**xamine the information given.
  - ☐ Reread the questions.
  - ☐ Underline key words or phrases.
  - ☐ Think about what the questions are asking.
- ☐ **S**ee if your answers match the questions.

64 COPYING IS PROHIBITED © Englefield & Associates, Inc.

Note: Student answers may vary. Example responses are for use as a guide.

Science on Target for Grade 3 Chapter 3—Activity 2

Activity 2

Step 1

Read the scenario "Plants Start It All!"

Plants Start It All!

The poster on the wall stated, "Plants Start It All!"

The poster illustrates a process called **photosynthesis**. This is the process that begins the food production for not only plants, but also for animals. Animals eat the plants to get the energy they need for life.

Plants are called **producers** because they make (produce) food using water, carbon dioxide, and energy from the sun.

Animals that eat plants or other animals are called **consumers**.

The animals that eat only plants are called **primary consumers** because they eat (consume) the plants as food.

Some animals eat other animals or organisms. They are called **secondary consumers** because they get the food produced by the plant from the animal or organism that ate it.

This is how energy is passed on to living organisms.

© Englefield & Associates, Inc. COPYING IS PROHIBITED 63

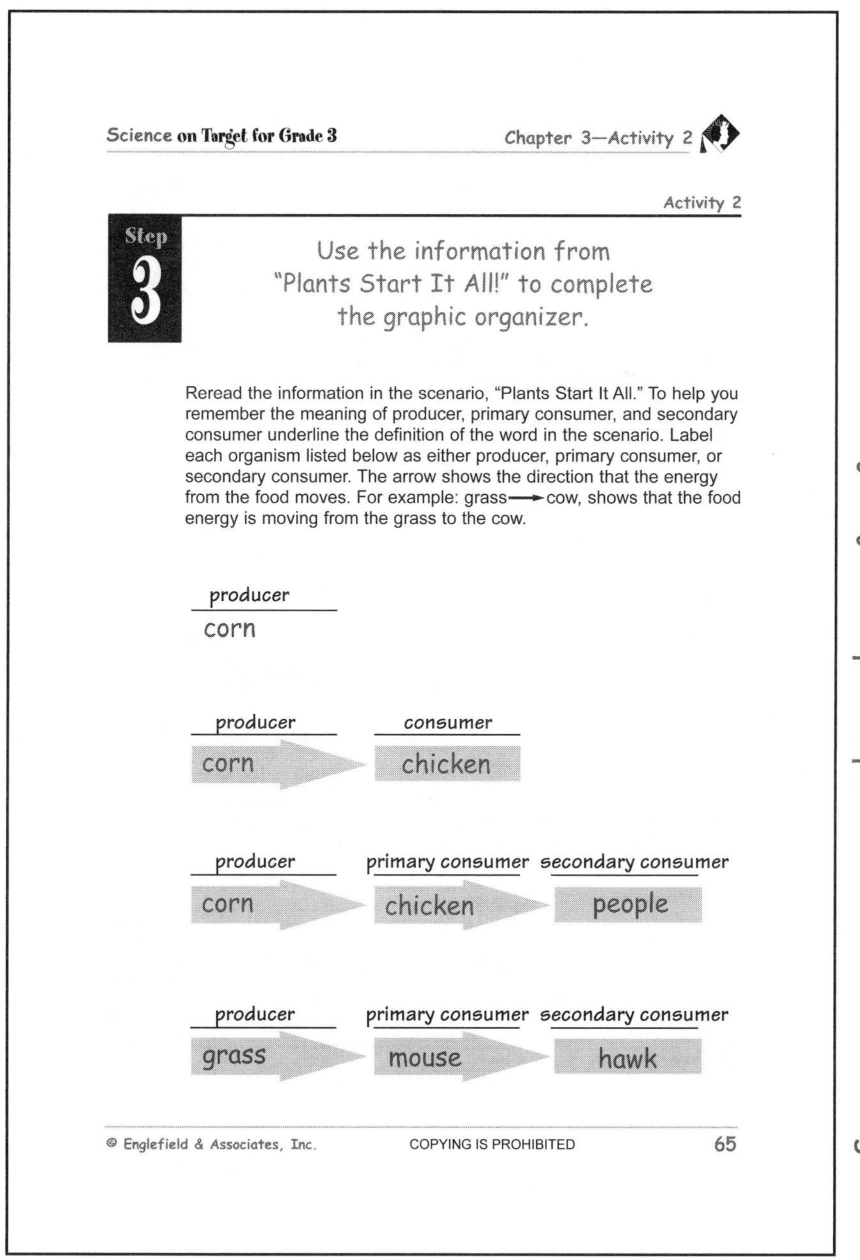

Step 3

Use the information from "Plants Start It All!" to complete the graphic organizer.

Reread the information in the scenario, "Plants Start It All." To help you remember the meaning of producer, primary consumer, and secondary consumer underline the definition of the word in the scenario. Label each organism listed below as either producer, primary consumer, or secondary consumer. The arrow shows the direction that the energy from the food moves. For example: grass ⟶ cow, shows that the food energy is moving from the grass to the cow.

Note: Student answers may vary. Example responses are for use as a guide.

Step 4

Answer the following questions for "Plants Start It All!" using information from your graphic organizer.

On a visit to the stream near their school, the students learned about the energy flow in water (aquatic) environments. In aquatic environments, plants capture the energy of the sun to start food production for living organisms.

Answer the questions below about the flow of energy in an aquatic environment.

1. Algae are plants that capture the energy of the sun to make food. Algae are

   producers (circled) or **consumers** (circle one).

2. Algae are eaten by zooplankton (very small animals that can be seen with a microscope).

   The **consumer** is zooplankton.

3. Waterstriders (insects that can walk on the surface of the water) eat zooplankton.

   The **primary consumer** is zooplankton.

   The **secondary consumer** is waterstrider.

4. Large mouth bass (fish) consume the waterstriders. The large mouth bass and waterstriders are both

   producers or **consumers** (circled) (circle one).

Activity 3

Step 1

Read the scenario "Surviving!"

Surviving!

An organism's pattern of behaviors are related to its environment. An organism may change when its environment changes. These changes are called **adaptations**. Animals that can adapt will survive and reproduce, while others may die or move to new locations. Physical adaptations may develop over many generations and will help the animal in several ways, include getting food, staying safe in the habitat, living in particular conditions, and attracting mates.

These adaptations fall into three main categories: body parts, body coverings, and behaviors. They are noticeable when you look at the animal or observe it over time.

The students researched several animals and their adaptations. Each student presented a three sentence summary about their organism's adaptation for the bulletin board.

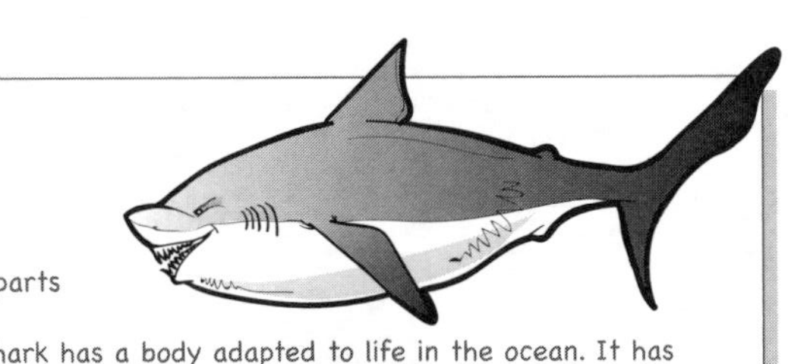

Body parts

The shark has a body adapted to life in the ocean. It has fins and a long, lean body that helps it swim. A shark breathes using gills that take oxygen out of the water to help it stay underwater for long periods of time.

Note: Student answers may vary. Example responses are for use as a guide.

Behavior

The raccoon lost much of its habitat to people. Raccoons have adapted by invading the spaces people created including their homes, garages, and storage sheds. With the loss of habitat, the raccoons' food sources became limited, so the raccoons learned to use their paws to open trash cans and steal food in campgrounds.

Body Coverings

The arctic fox has the ability to change the color of its coat to blend in with its environment. In the winter, the coat is white. In the summer, its coat is brown. This ability to camouflage itself helps the arctic fox sneak up on its prey and protects the arctic fox from its predators.

Body Coverings

The male peacock uses its bright feathers to attract a mate. The female has dull-colored feathers. This feature, common among female birds of most species, helps females remain unnoticed while they guard their nest to protect their young.

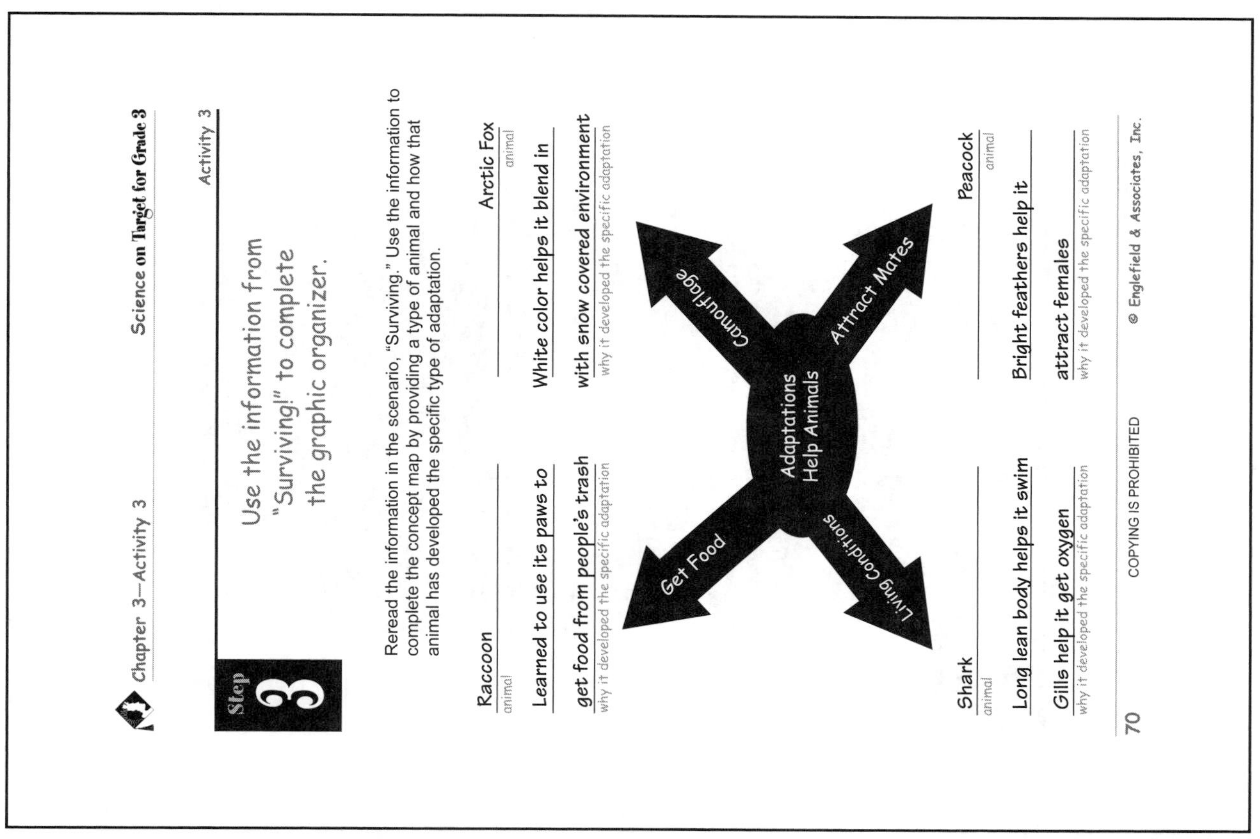

Chapter 3—Activity 3 Science on Target for Grade 3

Step 3

Use the information from "Surviving!" to complete the graphic organizer.

Activity 3

Reread the information in the scenario, "Surviving." Use the information to complete the concept map by providing a type of animal and how that animal has developed the specific type of adaptation.

Raccoon
animal

Learned to use its paws to
get food from people's trash
why it developed the specific adaptation

Arctic Fox
animal

White color helps it blend in
with snow covered environment
why it developed the specific adaptation

Shark
animal

Long lean body helps it swim
Gills help it get oxygen
why it developed the specific adaptation

Peacock
animal

Bright feathers help it
attract females
why it developed the specific adaptation

70 COPYING IS PROHIBITED © Englefield & Associates, Inc.

Note: Student answers may vary. Example responses are for use as a guide.

Science on Target for Grade 3 Chapter 3—Activity 3

Step 2

Complete the Checklist "Clues for Success."

The checklist will help you to read and think like a scientist.

Activity 3

Clues for Success

- ☐ **C**arefully read the information.
- ☐ **L**ook at any illustrations or diagrams.
  They may provide you with additional information to answer the question.
- ☐ **U**nderstand the way you are asked to answer the question.
  - ☐ Graph
  - ☐ Chart
  - ☐ Diagram
  - ☐ Complete sentences
  - ☐ Phrase
  - ☐ Filled circle
- ☐ **E**xamine the information given.
  - ☐ Reread the questions.
  - ☐ Underline key words or phrases.
  - ☐ Think about what the questions are asking.
- ☐ **S**ee if your answers match the questions.

© Englefield & Associates, Inc. COPYING IS PROHIBITED 69

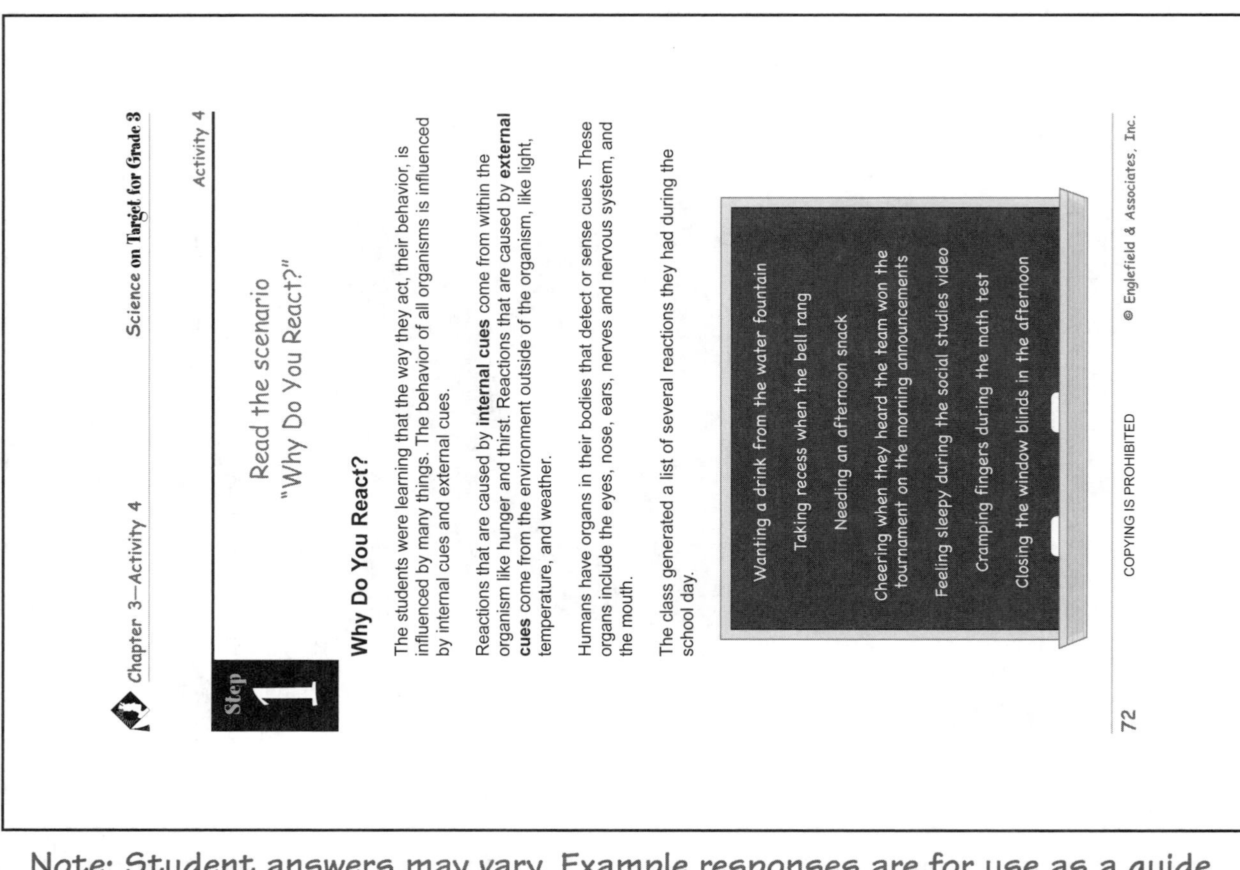

Chapter 3—Activity 4

Science on Target for Grade 3

Step 1

Activity 4

Read the scenario "Why Do You React?"

Why Do You React?

The students were learning that the way they act, their behavior, is influenced by many things. The behavior of all organisms is influenced by internal cues and external cues.

Reactions that are caused by **internal cues** come from within the organism like hunger and thirst. Reactions that are caused by **external cues** come from the environment outside of the organism, like light, temperature, and weather.

Humans have organs in their bodies that detect or sense cues. These organs include the eyes, nose, ears, nerves and nervous system, and the mouth.

The class generated a list of several reactions they had during the school day.

72 COPYING IS PROHIBITED © Englefield & Associates, Inc.

Note: Student answers may vary. Example responses are for use as a guide.

Science on Target for Grade 3

Chapter 3—Activity 3

Step 4

Activity 3

Answer the following questions for "Surviving!" using information from your graphic organizer.

Consider the animals and the adaptations listed below. Use a complete sentence to explain how the adaptation helps each animal.

1. Bears hibernate. Bears hibernate to survive in the winter.
2. Ducks have webbed feet. Webbed feet help ducks swim in water.
3. Giraffes have long necks. Long necks help giraffes reach food.
4. Porcupines have quills. Quills help protect porcupines from predators in their environment.

© Englefield & Associates, Inc. COPYING IS PROHIBITED 71

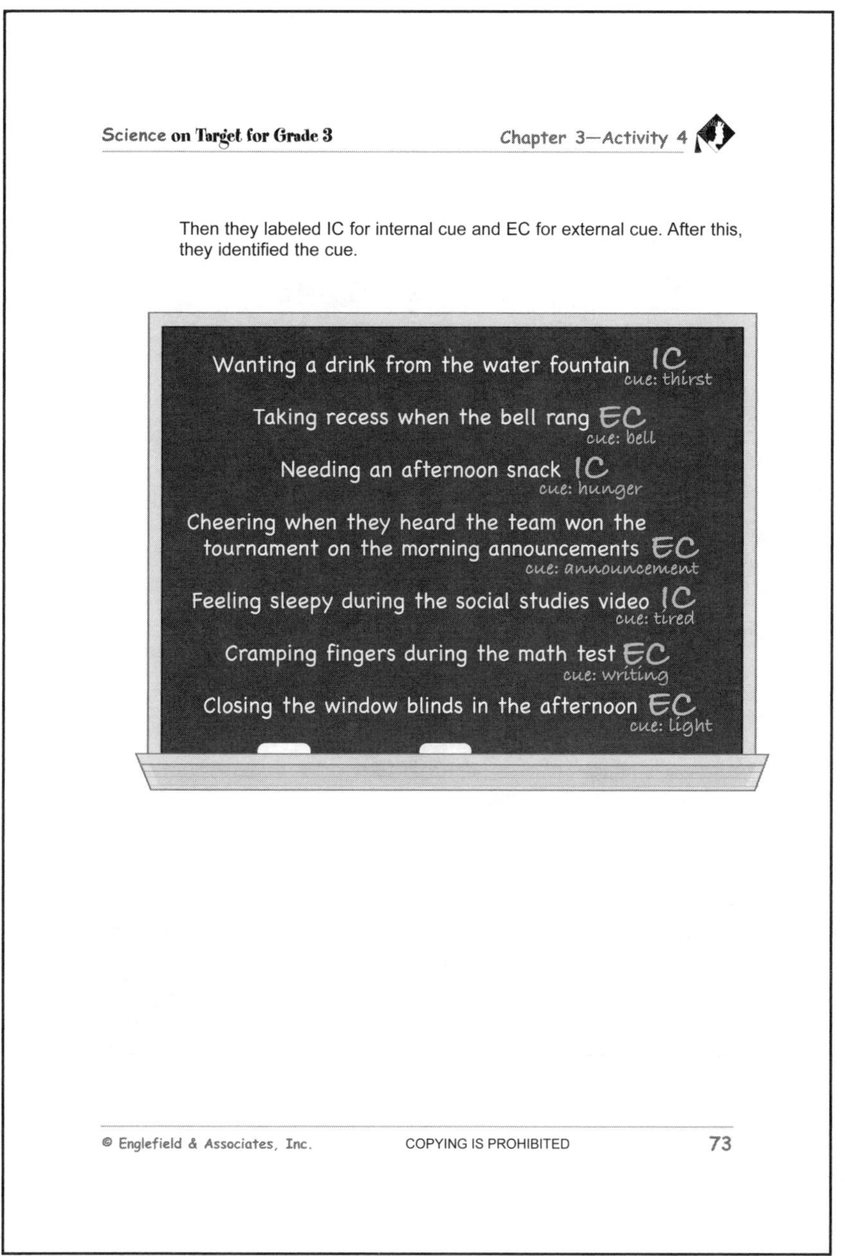

Science on Target for Grade 3 Chapter 3—Activity 4

Then they labeled IC for internal cue and EC for external cue. After this, they identified the cue.

© Englefield & Associates, Inc. COPYING IS PROHIBITED 73

Note: Student answers may vary. Example responses are for use as a guide.

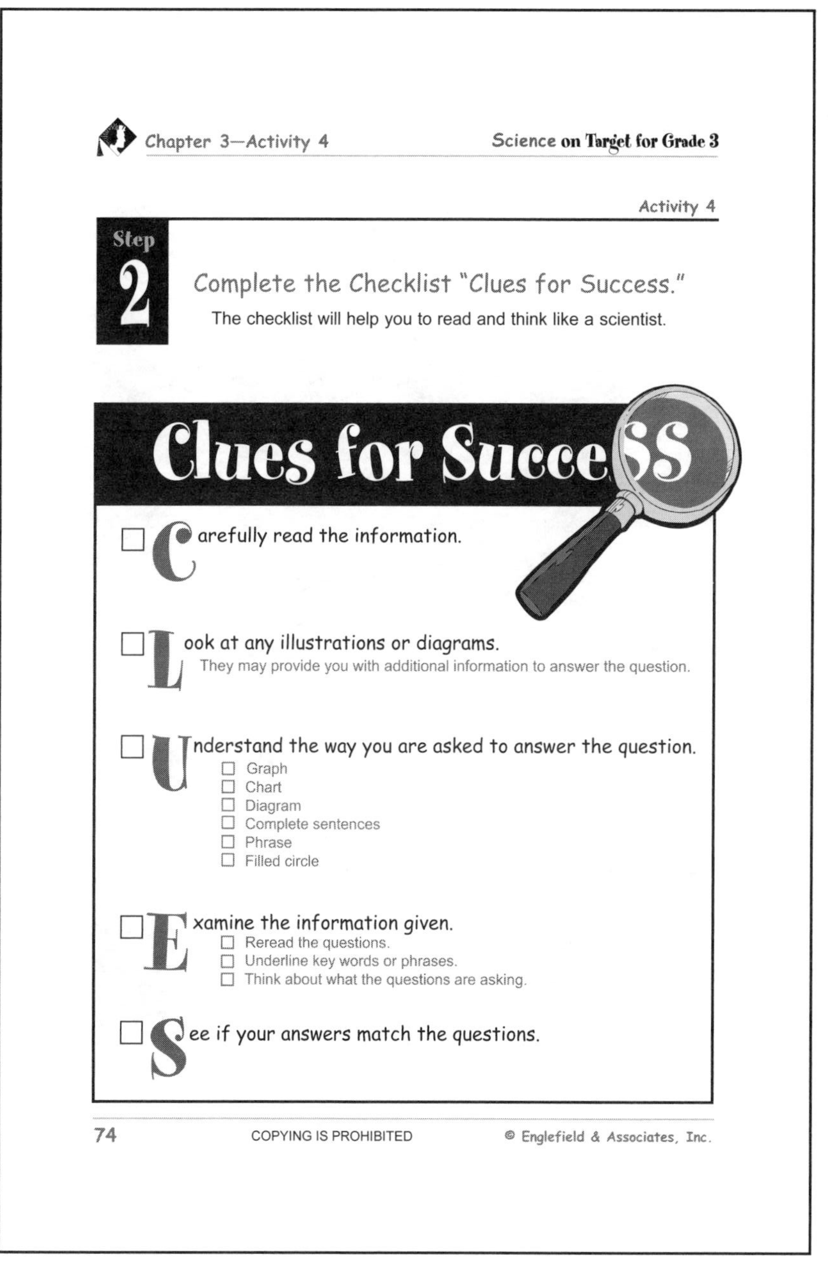

Chapter 3—Activity 4 Science on Target for Grade 3

Step 2 Activity 4

Complete the Checklist "Clues for Success."

The checklist will help you to read and think like a scientist.

Clues for Success

- ☐ **C**arefully read the information.
- ☐ **L**ook at any illustrations or diagrams.
  They may provide you with additional information to answer the question.
- ☐ **U**nderstand the way you are asked to answer the question.
  - ☐ Graph
  - ☐ Chart
  - ☐ Diagram
  - ☐ Complete sentences
  - ☐ Phrase
  - ☐ Filled circle
- ☐ **E**xamine the information given.
  - ☐ Reread the questions.
  - ☐ Underline key words or phrases.
  - ☐ Think about what the questions are asking.
- ☐ **S**ee if your answers match the questions.

74 COPYING IS PROHIBITED © Englefield & Associates, Inc.

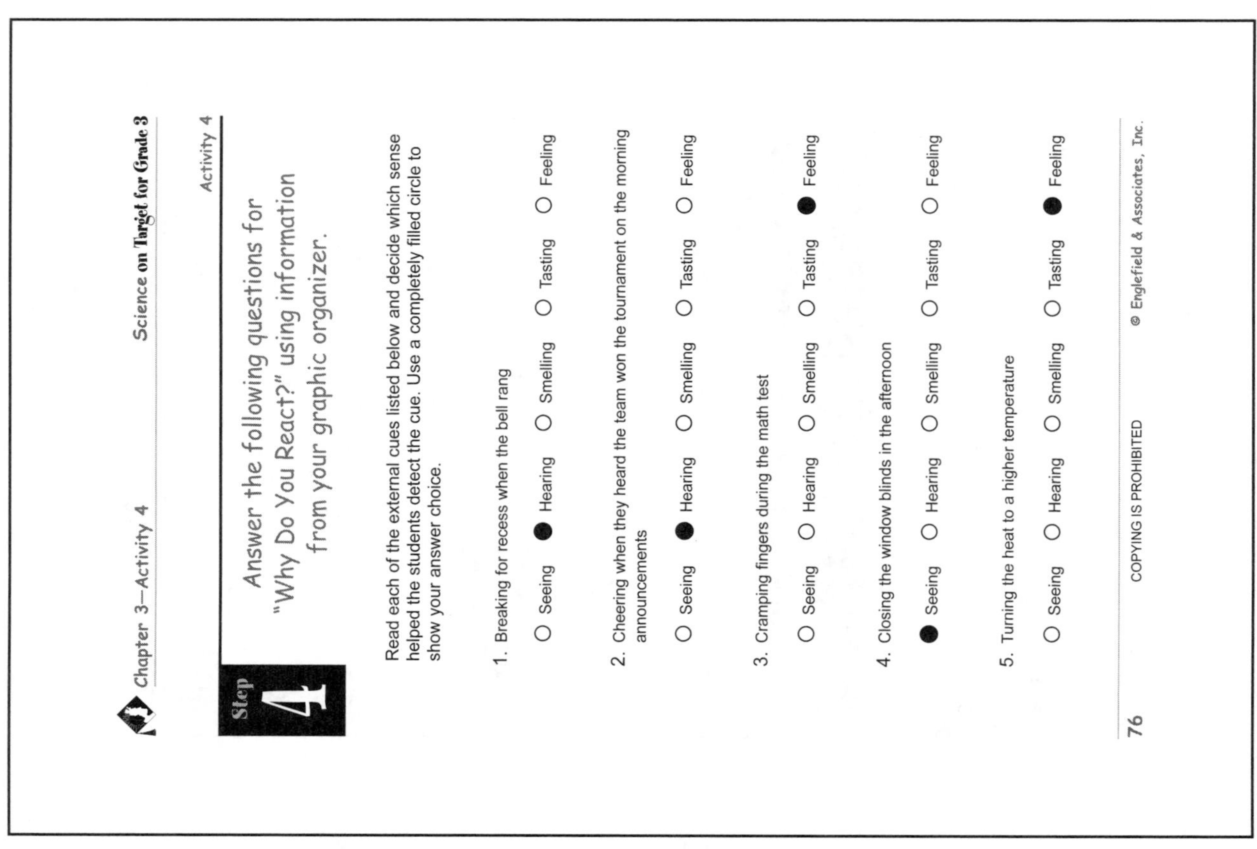

Activity 4

Step 4

Answer the following questions for "Why Do You React?" using information from your graphic organizer.

Read each of the external cues listed below and decide which sense helped the students detect the cue. Use a completely filled circle to show your answer choice.

1. Breaking for recess when the bell rang
   - ○ Seeing ● Hearing ○ Smelling ○ Tasting ○ Feeling
2. Cheering when they heard the team won the tournament on the morning announcements
   - ○ Seeing ● Hearing ○ Smelling ○ Tasting ○ Feeling
3. Cramping fingers during the math test
   - ○ Seeing ○ Hearing ○ Smelling ○ Tasting ● Feeling
4. Closing the window blinds in the afternoon
   - ● Seeing ○ Hearing ○ Smelling ○ Tasting ○ Feeling
5. Turning the heat to a higher temperature
   - ○ Seeing ○ Hearing ○ Smelling ○ Tasting ● Feeling

Note: Student answers may vary. Example responses are for use as a guide.

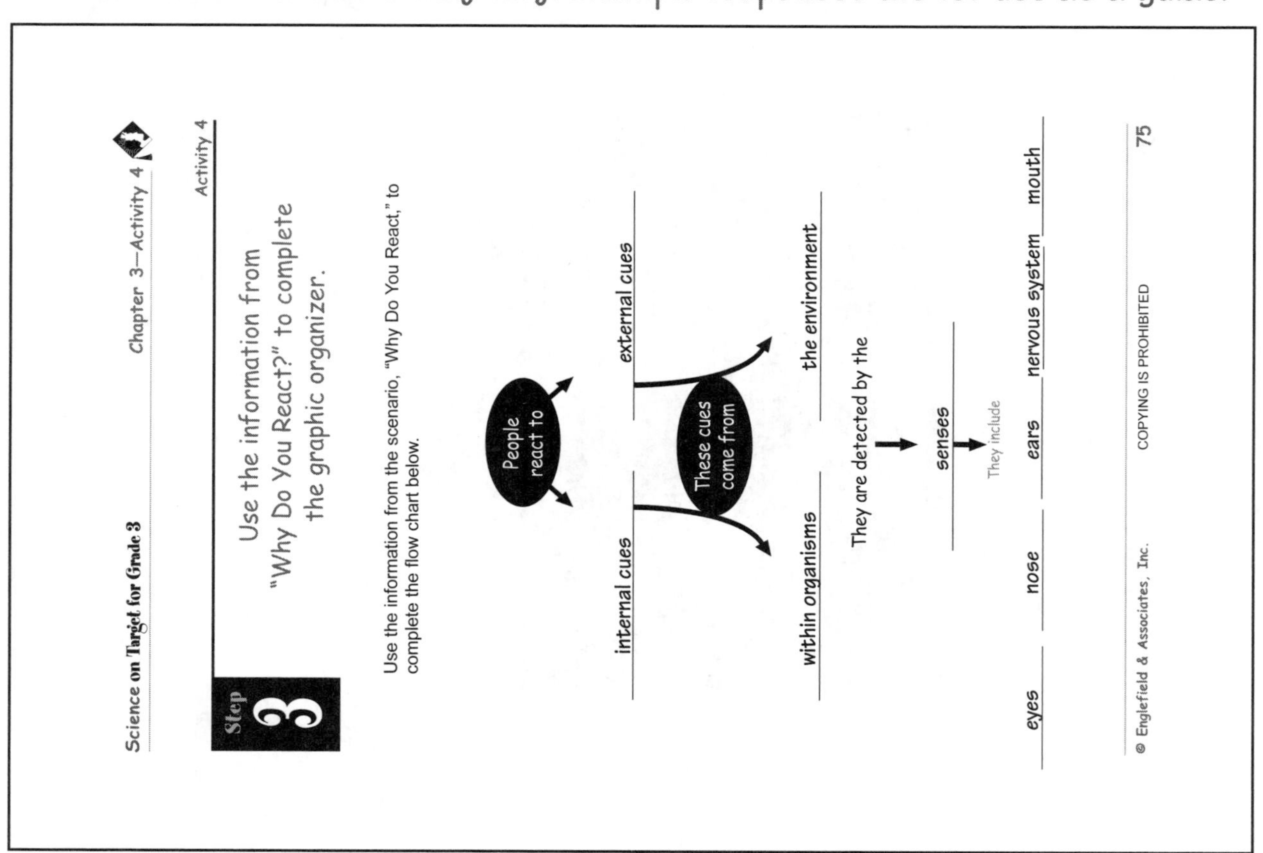

Activity 4

Step 3

Use the information from "Why Do You React?" to complete the graphic organizer.

Use the information from the scenario, "Why Do You React," to complete the flow chart below.

Chapter 3—Activity 5 Science on Target for Grade 3

Activity 5

Step 1

Read the scenario "Life Cycles of Organisms."

Life Cycles of Organisms

The students learned about the life cycle of organisms as they watched a video clip showing the transformation of an egg into a butterfly. They learned that butterflies, like all living organisms, have life cycles that included a variety of stages including being born, developing into adults, reproducing, and eventually dying. The butterfly is an example of an insect that undergoes complete metamorphosis.

Students created a diagram to help them learn the stages in the life cycle of a butterfly.

78 COPYING IS PROHIBITED © Englefield & Associates, Inc.

Note: Student answers may vary. Example responses are for use as a guide.

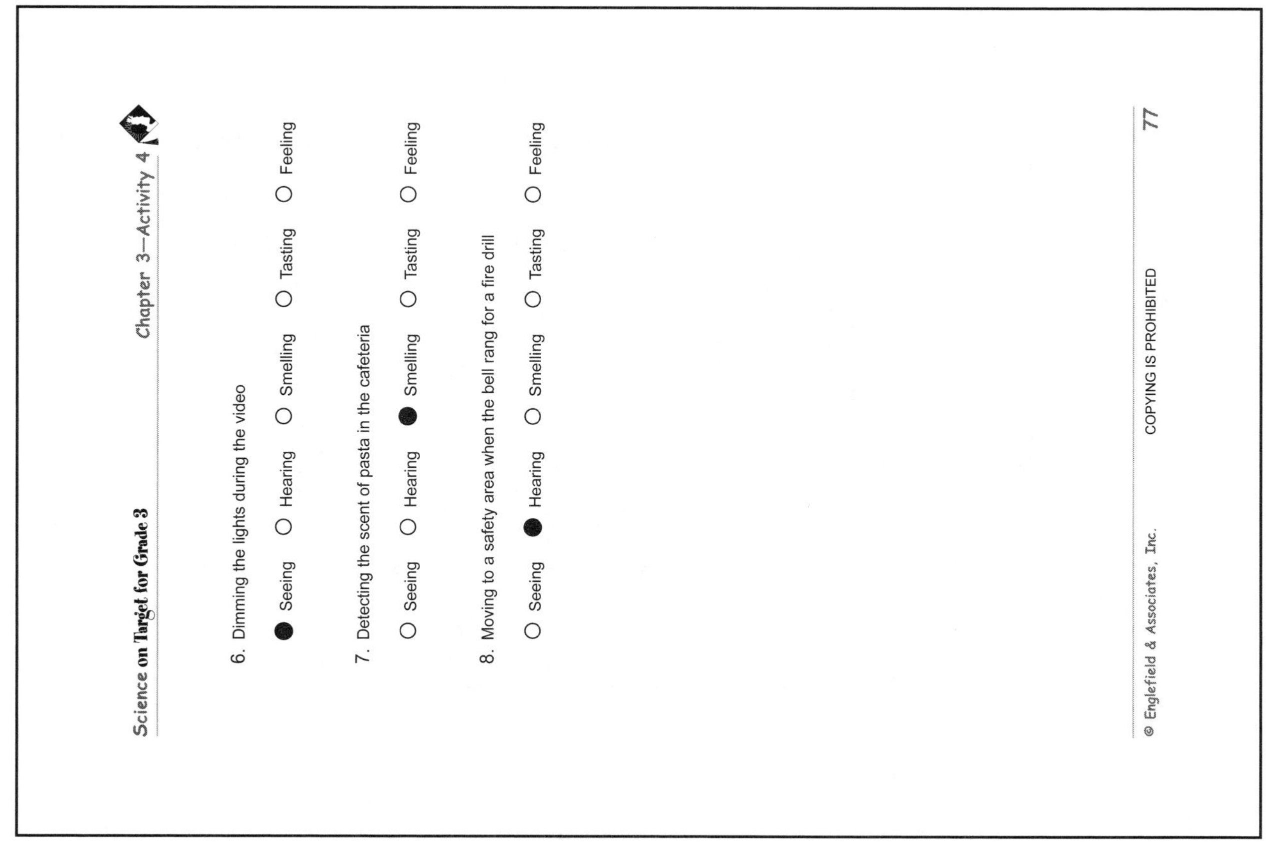

Science on Target for Grade 3 Chapter 3—Activity 4

6. Dimming the lights during the video

● Seeing ○ Hearing ○ Smelling ○ Tasting ○ Feeling

7. Detecting the scent of pasta in the cafeteria

○ Seeing ○ Hearing ● Smelling ○ Tasting ○ Feeling

8. Moving to a safety area when the bell rang for a fire drill

○ Seeing ● Hearing ○ Smelling ○ Tasting ○ Feeling

© Englefield & Associates, Inc. COPYING IS PROHIBITED 77

Chapter 3—Activity 5 Science on Target for Grade 3

Step 2 Activity 5

Complete the Checklist "Clues for Success."

The checklist will help you to read and think like a scientist.

Clues for Success

- ☐ **C**arefully read the information.
- ☐ **L**ook at any illustrations or diagrams.
  They may provide you with additional information to answer the question.
- ☐ **U**nderstand the way you are asked to answer the question.
  - ☐ Graph
  - ☐ Chart
  - ☐ Diagram
  - ☐ Complete sentences
  - ☐ Phrase
  - ☐ Filled circle
- ☐ **E**xamine the information given.
  - ☐ Reread the questions.
  - ☐ Underline key words or phrases.
  - ☐ Think about what the questions are asking.
- ☐ **S**ee if your answers match the questions.

80 COPYING IS PROHIBITED © Englefield & Associates, Inc.

Note: Student answers may vary. Example responses are for use as a guide.

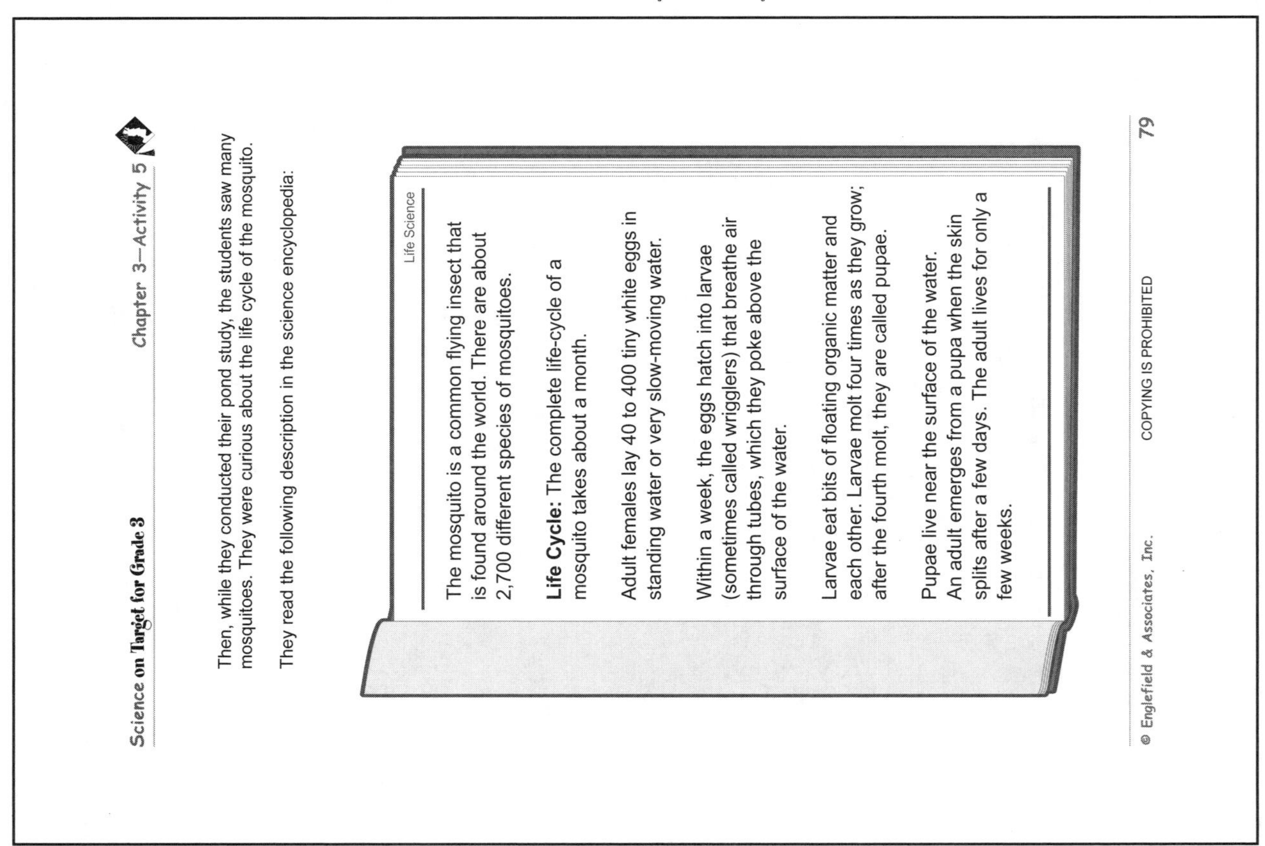
Science on Target for Grade 3 Chapter 3—Activity 5

Then, while they conducted their pond study, the students saw many mosquitoes. They were curious about the life cycle of the mosquito.

They read the following description in the science encyclopedia:

Life Science

The mosquito is a common flying insect that is found around the world. There are about 2,700 different species of mosquitoes.

Life Cycle: The complete life-cycle of a mosquito takes about a month.

Adult females lay 40 to 400 tiny white eggs in standing water or very slow-moving water.

Within a week, the eggs hatch into larvae (sometimes called wrigglers) that breathe air through tubes, which they poke above the surface of the water.

Larvae eat bits of floating organic matter and each other. Larvae molt four times as they grow; after the fourth molt, they are called pupae.

Pupae live near the surface of the water. An adult emerges from a pupa when the skin splits after a few days. The adult lives for only a few weeks.

© Englefield & Associates, Inc. COPYING IS PROHIBITED 79

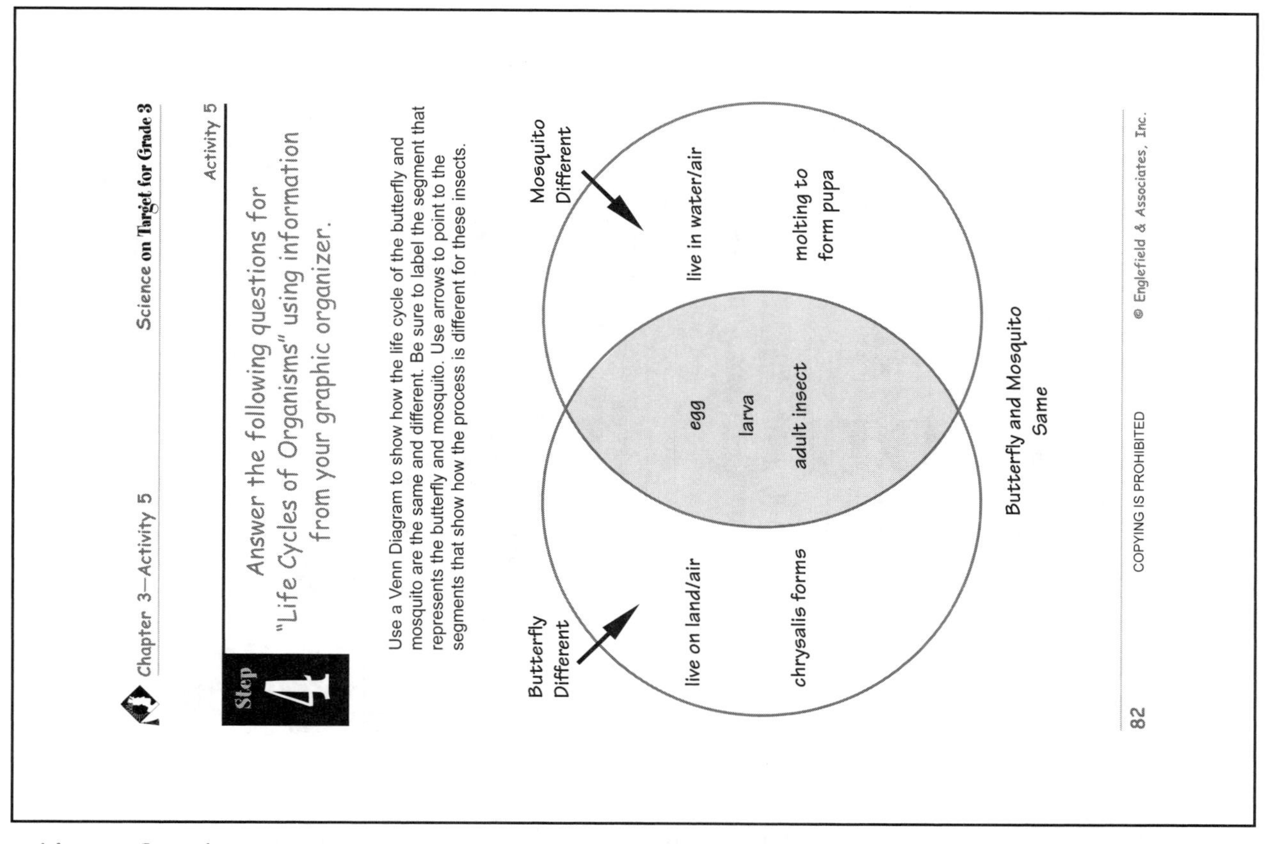

Step 4

Answer the following questions for "Life Cycles of Organisms" using information from your graphic organizer.

Activity 5

Use a Venn Diagram to show how the life cycle of the butterfly and mosquito are the same and different. Be sure to label the segment that represents the butterfly and mosquito. Use arrows to point to the segments that show how the process is different for these insects.

Note: Student answers may vary. Example responses are for use as a guide.

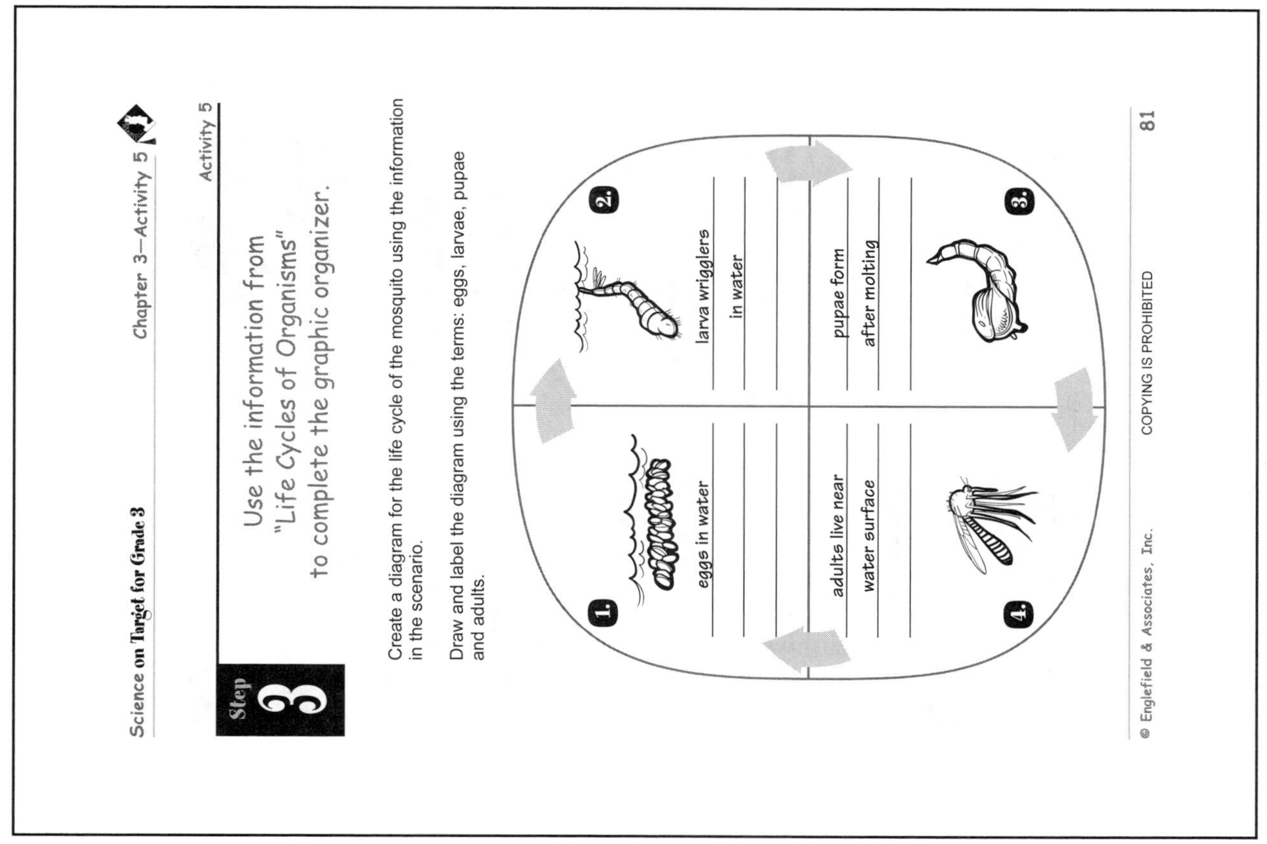

Step 3

Use the information from "Life Cycles of Organisms" to complete the graphic organizer.

Activity 5

Create a diagram for the life cycle of the mosquito using the information in the scenario.

Draw and label the diagram using the terms: eggs, larvae, pupae and adults.

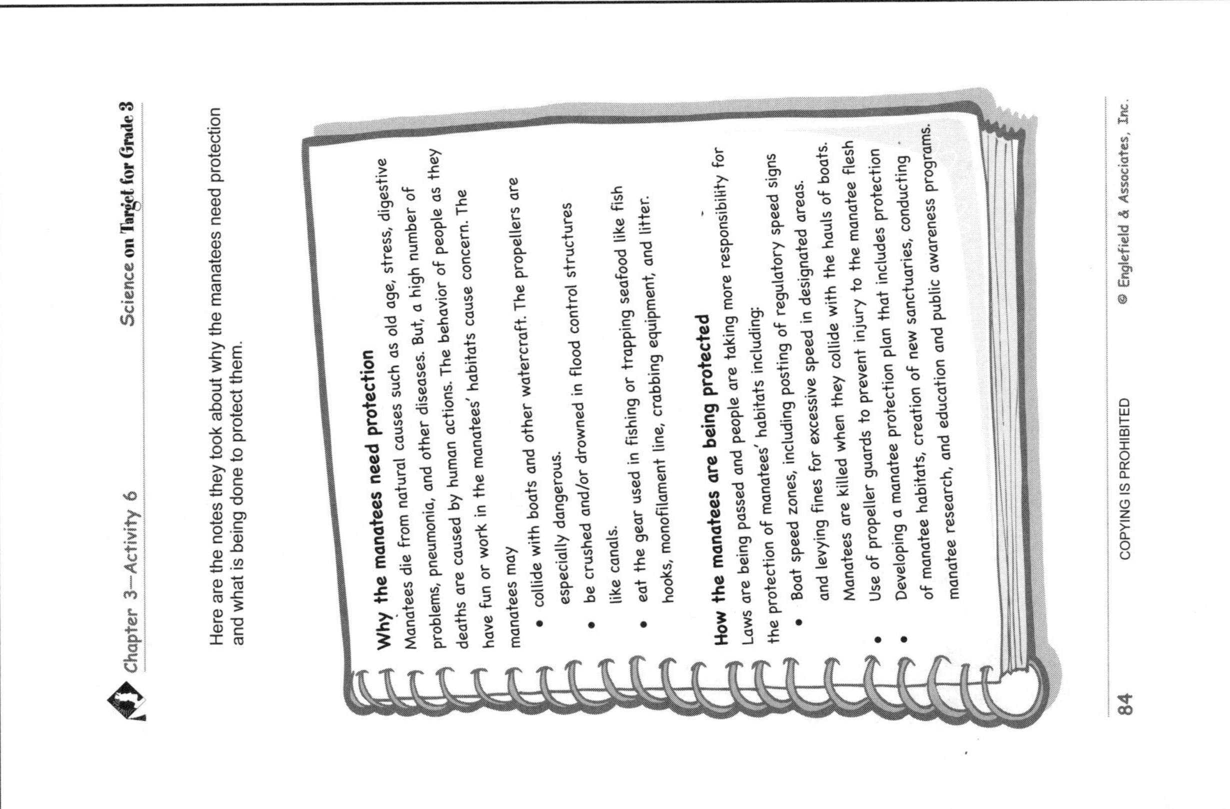

Here are the notes they took about why the manatees need protection and what is being done to protect them.

Why the manatees need protection

Manatees die from natural causes such as old age, stress, digestive problems, pneumonia, and other diseases. But, a high number of deaths are caused by human actions. The behavior of people as they have fun or work in the manatees' habitats cause concern. The manatees may

- collide with boats and other watercraft. The propellers are especially dangerous.
- be crushed and/or drowned in flood control structures like canals.
- eat the gear used in fishing or trapping seafood like fish hooks, monofilament line, crabbing equipment, and litter.

How the manatees are being protected

Laws are being passed and people are taking more responsibility for the protection of manatees' habitats including:

- Boat speed zones, including posting of regulatory speed signs and levying fines for excessive speed in designated areas. Manatees are killed when they collide with the hauls of boats.
- Use of propeller guards to prevent injury to the manatee flesh
- Developing a manatee protection plan that includes protection of manatee habitats, creation of new sanctuaries, conducting manatee research, and education and public awareness programs.

Note: Student answers may vary. Example responses are for use as a guide.

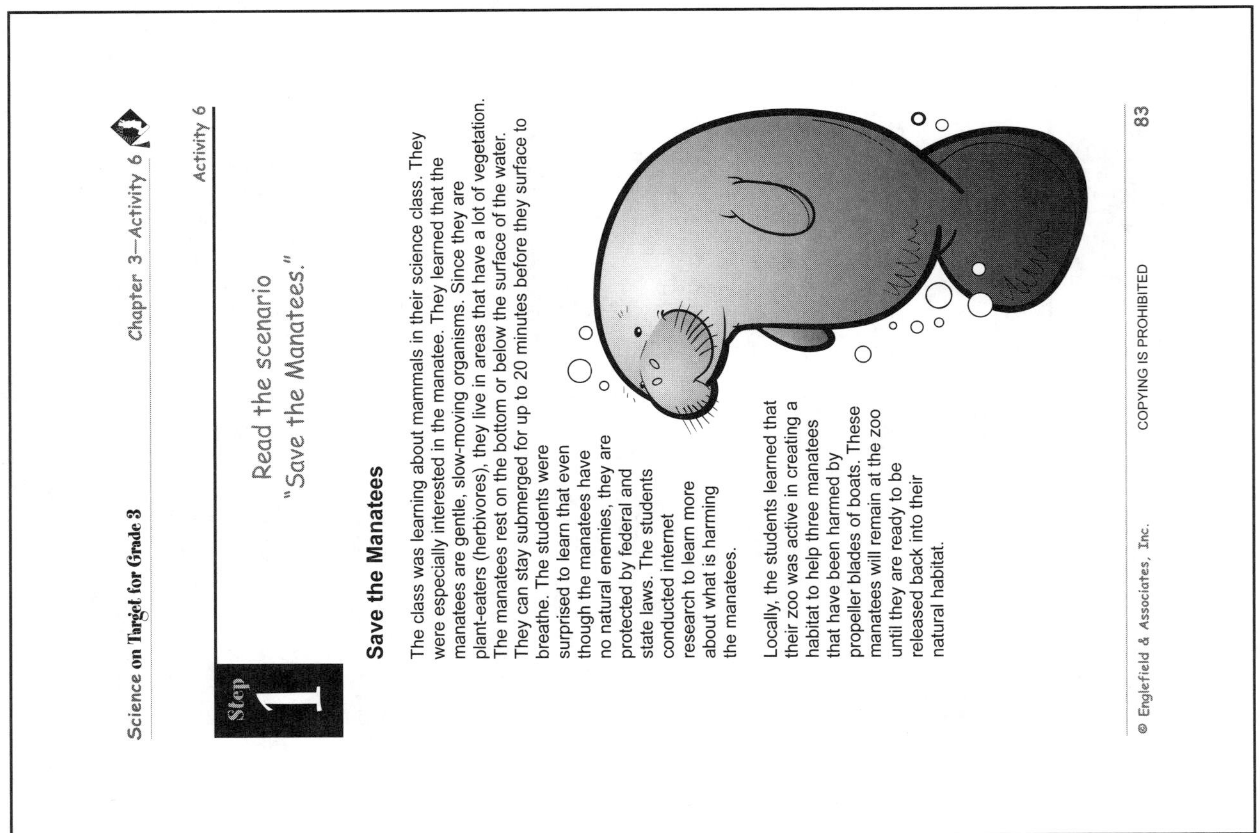

Activity 6

Step 1

Read the scenario "Save the Manatees."

Save the Manatees

The class was learning about mammals in their science class. They were especially interested in the manatee. They learned that the manatees are gentle, slow-moving organisms. Since they are plant-eaters (herbivores), they live in areas that have a lot of vegetation. The manatees rest on the bottom or below the surface of the water. They can stay submerged for up to 20 minutes before they surface to breathe. The students were surprised to learn that even though the manatees have no natural enemies, they are protected by federal and state laws. The students conducted internet research to learn more about what is harming the manatees.

Locally, the students learned that their zoo was active in creating a habitat to help three manatees that have been harmed by propeller blades of boats. These manatees will remain at the zoo until they are ready to be released back into their natural habitat.

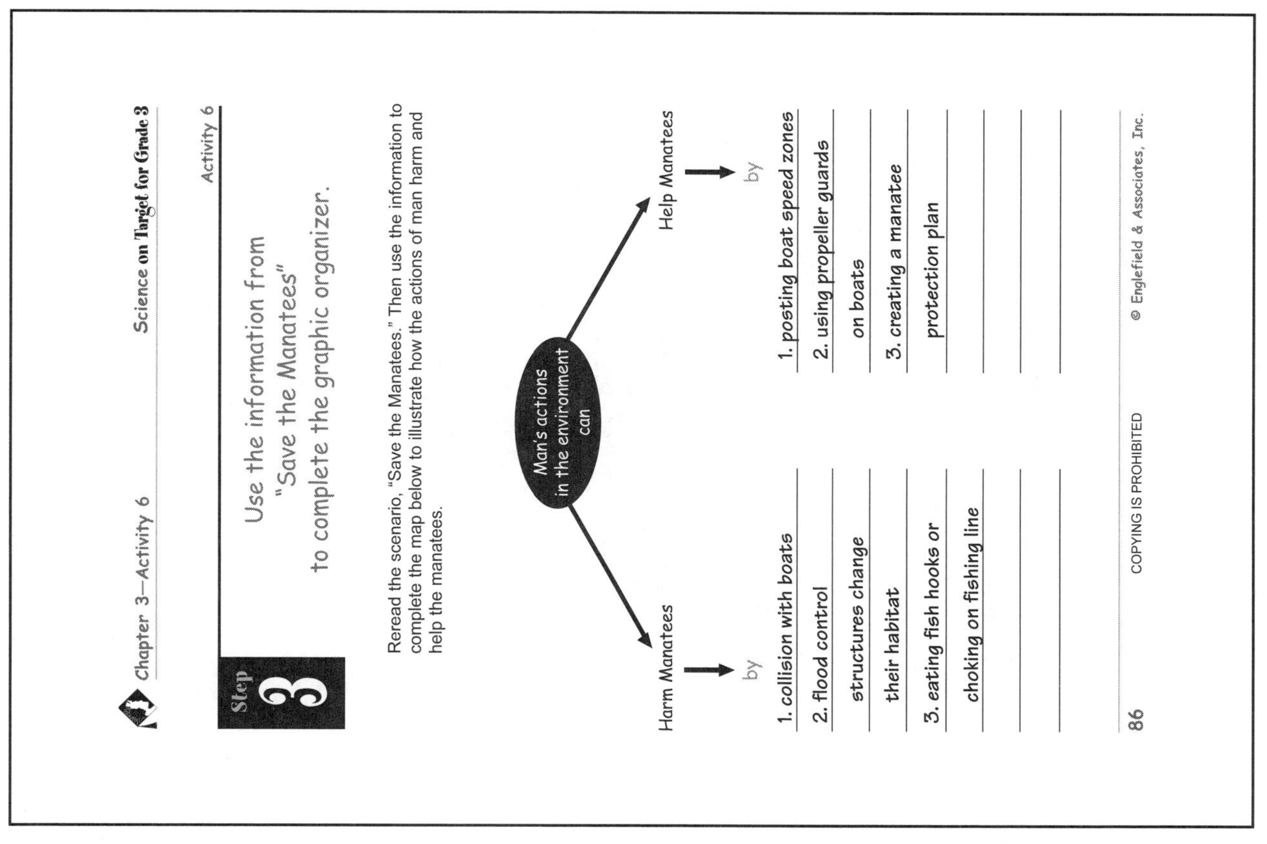
Chapter 3—Activity 6 Science on Target for Grade 3

Activity 6

Step 3

Use the information from "Save the Manatees" to complete the graphic organizer.

Reread the scenario, "Save the Manatees." Then use the information to complete the map below to illustrate how the actions of man harm and help the manatees.

Man's actions in the environment can

Harm Manatees by	Help Manatees by
1. collision with boats	1. posting boat speed zones
2. flood control structures change their habitat	2. using propeller guards on boats
3. eating fish hooks or choking on fishing line	3. creating a manatee protection plan

86 COPYING IS PROHIBITED © Englefield & Associates, Inc.

Note: Student answers may vary. Example responses are for use as a guide.

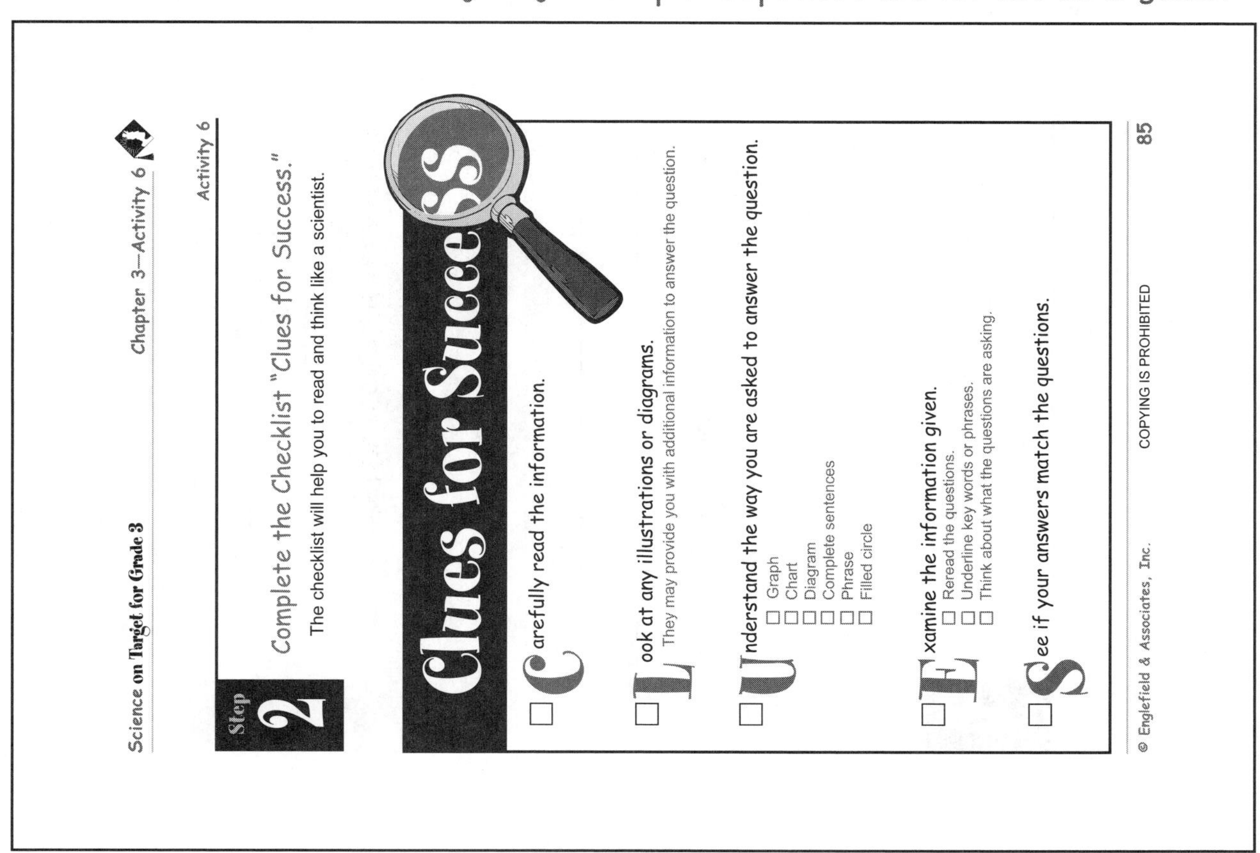
Science on Target for Grade 3 Chapter 3—Activity 6

Activity 6

Step 2

Complete the Checklist "Clues for Success."

The checklist will help you to read and think like a scientist.

Clues for Success

- ☐ **C**arefully read the information.
- ☐ **L**ook at any illustrations or diagrams. They may provide you with additional information to answer the question.
- ☐ **U**nderstand the way you are asked to answer the question.
  - ☐ Graph
  - ☐ Chart
  - ☐ Diagram
  - ☐ Complete sentences
  - ☐ Phrase
  - ☐ Filled circle
- ☐ **E**xamine the information given.
  - ☐ Reread the questions.
  - ☐ Underline key words or phrases.
  - ☐ Think about what the questions are asking.
- ☐ **S**ee if your answers match the questions.

© Englefield & Associates, Inc. COPYING IS PROHIBITED 85

Activity 6

Step 4

Answer the following questions for "Save the Manatees" using information from your graphic organizer.

In the space provided, use a complete sentence to show how people's behavior impacts the manatees.

1. People protect their environment from flooding by using flood walls.

   When people use flood walls, the amount of water in the manatees environment is changed.

2. People are having fun boating in a manatee habitat.

   If people are using boats with propellers in a manatee habitat, the propellers can cut the manatee's body.

3. People are working as fisherman in a manatee habitat.

   The equipment the people use to fish may be eaten by the manatees and harm them.

4. Zoo keepers provide a safe home for injured manatees.

   When zookeepers provide a safe home, the manatees can safely recover from their injuries.

Note: Student answers may vary. Example responses are for use as a guide.

Chapter 4

Chapter 4 of the *Science on Target for Grade 3 Student Workbook*, covers the National Content Standards for Earth Science. The standards are as follows:

Earth Science

Earth Science

Properties of Earth materials

- Earth materials are solid rocks and soils, water, and the gases of the atmosphere. The varied materials have different physical and chemical properties which make them useful in different ways, for example, as building materials, as sources of fuel, or for growing the plants we use as food. Earth materials provide many of the resources that humans use.
- Soils have properties of color and texture, capacity to retain water, and ability to support the growth of many kinds of plants, including those in our food supply.
- Fossils provide evidence about the plants and animals that lived long ago and the nature of the environment at that time.

Objects in the sky

- The sun, moon, stars, clouds, birds, and airplanes all have properties, locations, and movements that can be observed and described.
- The sun provides the light and heat necessary to maintain the temperature of Earth.

Changes in Earth and the sky

- The surface of Earth changes. Some changes are due to slow processes, such as erosion and weathering, and some changes are due to rapid processes, such as landslides, volcanic eruptions, and earthquakes.
- Weather changes from day to day and over seasons. Weather can be described by measurable quantities, such as temperature, wind direction and speed, and precipitation.

All of the pages from Chapter 4 of the *Science on Target for Grade 3 Student Workbook*, are reproduced in this Parent/Teacher Edition in reduced-page format with sample answers. These activities will help your students develop the skills necessary to understand Earth Science.

Students should use the "Clues for Success" Checklis,t for each activity in this section, as a tool to help them do their best work.

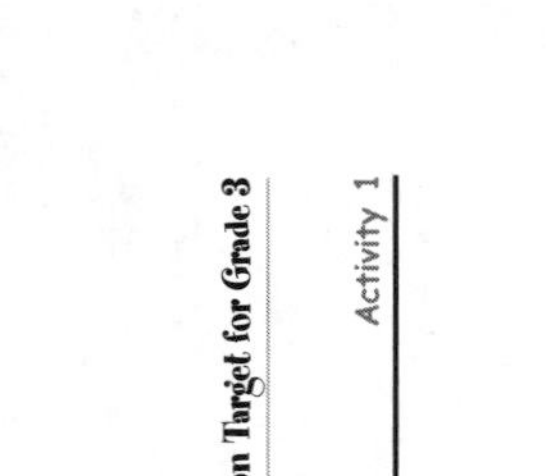

Step 1

Read the scenario "Gifts of Earth."

Gifts of Earth

Earth provides materials for the needs of all living things. These materials are useful in many different ways. Generally, the gifts of Earth come from three areas: the lithosphere, the atmosphere, and the hydrosphere.

The materials provided from the lithosphere include materials from Earth's crust, such as solid rocks, soil, and sand. They also include those machine-obtained materials from the earth, such as fuels (coal, oil, and natural gas) and minerals (the elements of gold and silver).

The atmosphere provides the gases used and exchanged by living organisms, such as oxygen (needed by animals) and carbon dioxide (needed by plants). The atmosphere also includes the layers of gases that protect Earth, like the ozone layer, and those that build up in the atmosphere that may be harmful, like the greenhouse gases.

Water (necessary for life) is found in the hydrosphere. Water can be found in different states, such as:

- **liquid** in the oceans or rain,
- **solid** in the ice caps at the poles or the icebergs in the arctic, snow, or hail, and
- **gas** or **vapor** in the clouds or fog.

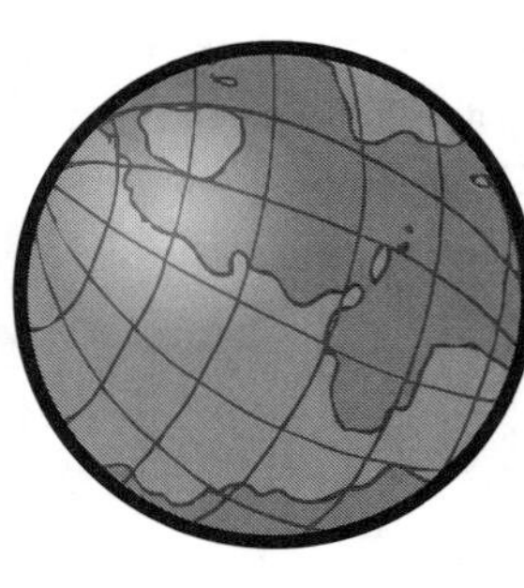

Note: Student answers may vary. Example responses are for use as a guide.

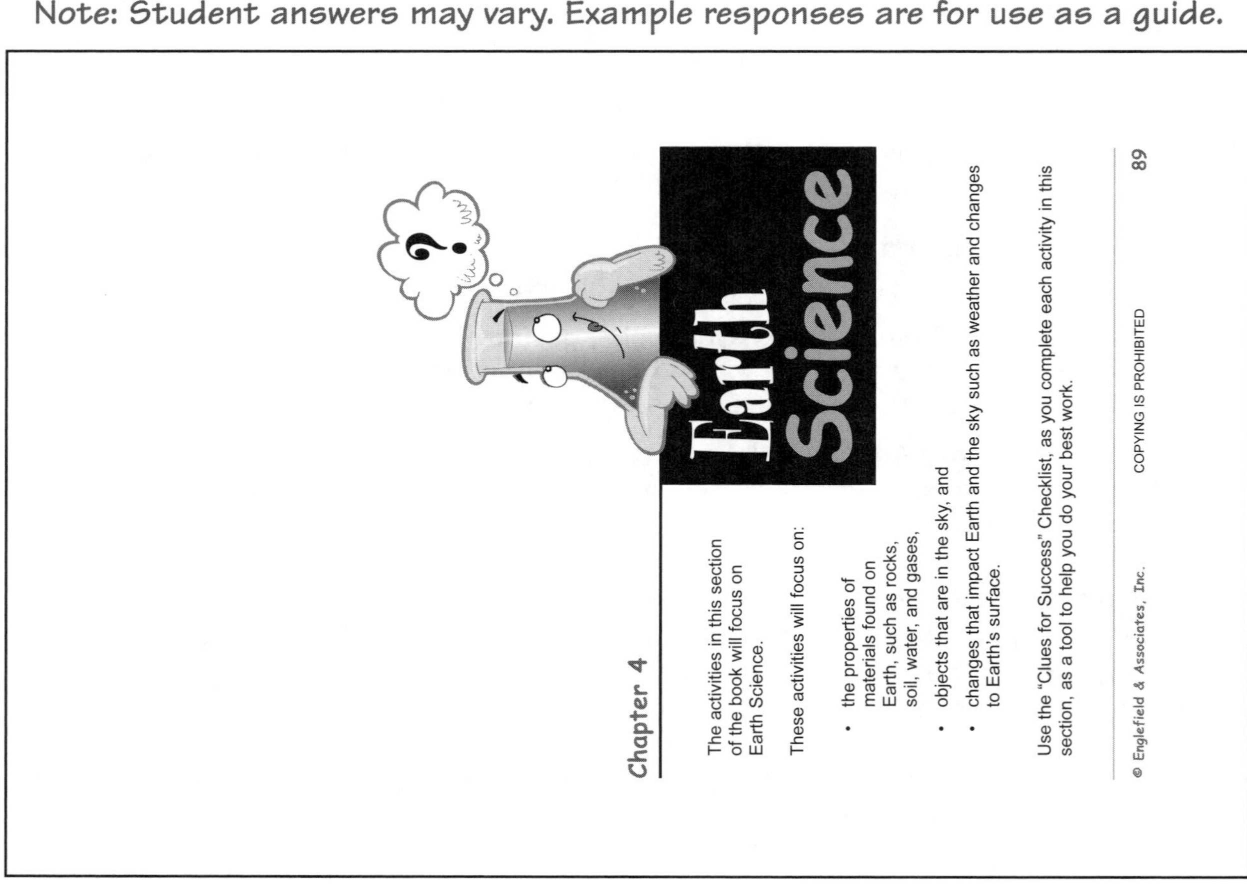

Chapter 4

Earth Science

The activities in this section of the book will focus on Earth Science.

These activities will focus on:

- the properties of materials found on Earth, such as rocks, soil, water, and gases,
- objects that are in the sky, and
- changes that impact Earth and the sky such as weather and changes to Earth's surface.

Use the "Clues for Success" Checklist, as you complete each activity in this section, as a tool to help you do your best work.

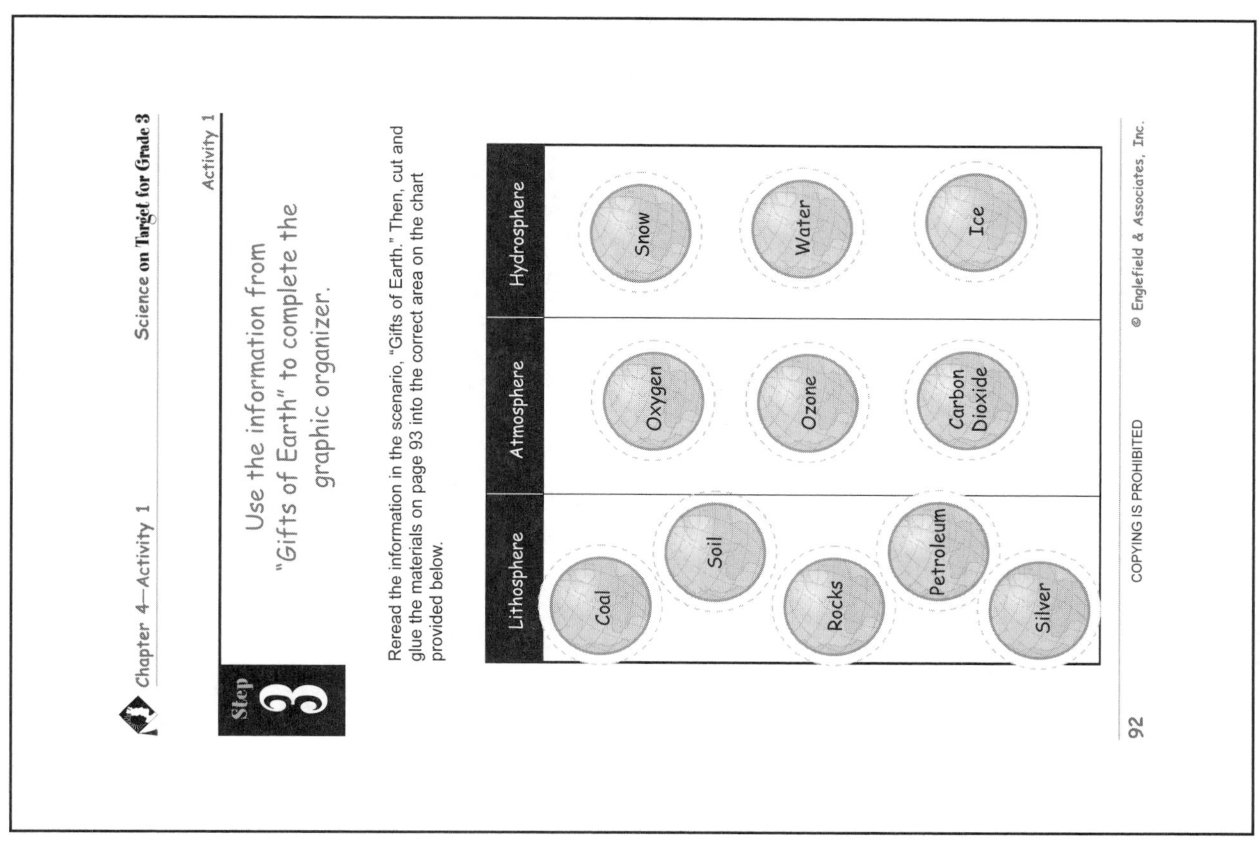

Step 3

Use the information from "Gifts of Earth" to complete the graphic organizer.

Reread the information in the scenario, "Gifts of Earth." Then, cut and glue the materials on page 93 into the correct area on the chart provided below.

Lithosphere	Atmosphere	Hydrosphere
Coal, Soil, Rocks, Petroleum, Silver	Oxygen, Ozone, Carbon Dioxide	Snow, Water, Ice

Note: Student answers may vary. Example responses are for use as a guide.

Step 2

Complete the Checklist "Clues for Success."

The checklist will help you to read and think like a scientist.

Clues for Success

- ☐ **C**arefully read the information.
- ☐ **L**ook at any illustrations or diagrams.
  They may provide you with additional information to answer the question.
- ☐ **U**nderstand the way you are asked to answer the question.
  - ☐ Graph
  - ☐ Chart
  - ☐ Diagram
  - ☐ Complete sentences
  - ☐ Phrase
  - ☐ Filled circle
- ☐ **E**xamine the information given.
  - ☐ Reread the questions.
  - ☐ Underline key words or phrases.
  - ☐ Think about what the questions are asking.
- ☐ **S**ee if your answers match the questions.

Note: Student answers may vary. Example responses are for use as a guide.

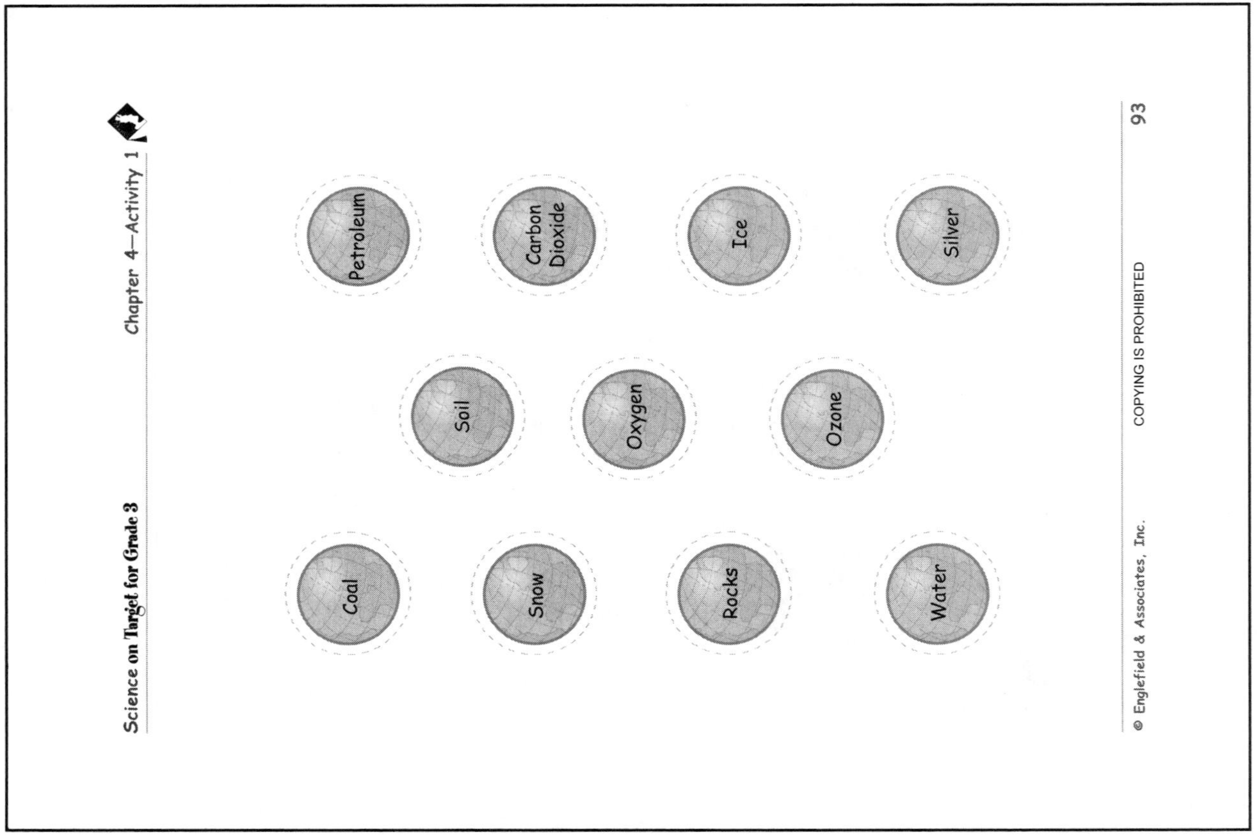

Note: Student answers may vary. Example responses are for use as a guide.

Step 4 — Activity 1

Answer the following questions for "Gifts of Earth" using information from your graphic organizer.

Plants use materials from each sphere of Earth to live. Identify a material from each sphere that the plants use. Then use a complete sentence to tell how the plants use the material to live.

1. **Lithosphere**

   Material used:

   soil

   How is the material used?

   Plants have roots in the soil. The roots keep plants in place and help provide water and nutrients.

2. **Atmosphere**

   Material used:

   carbon dioxide

   How is the material used?

   Plants combine carbon dioxide and water using energy from the sun to make food (sugar).

3. **Hydrosphere**

   Material used:

   water

   How is the material used?

   Plants combine water and carbon dioxide using the energy from the sun to make food (sugar).

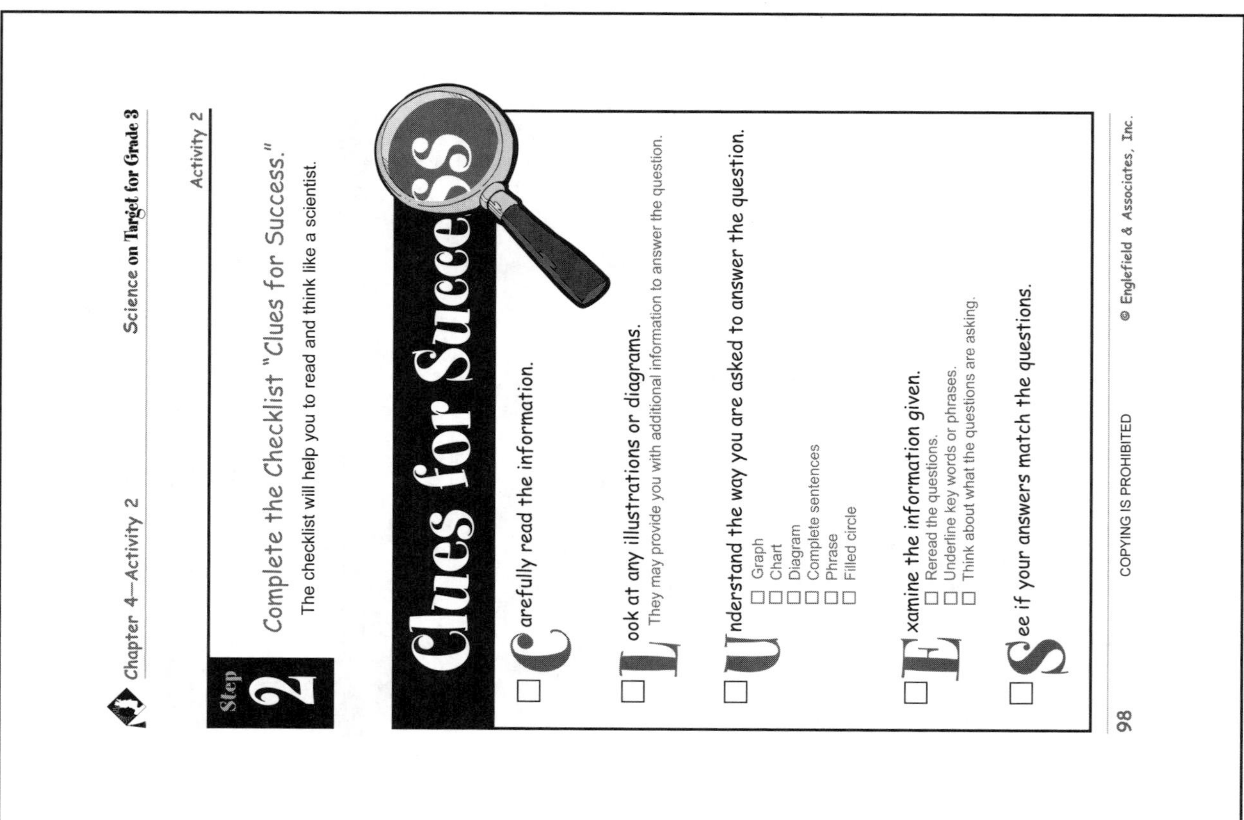

Chapter 4—Activity 2 Science on Target for Grade 3

Activity 2

Step 2 Complete the Checklist "Clues for Success."

The checklist will help you to read and think like a scientist.

Clues for Success

- ☐ **C**arefully read the information.
- ☐ **L**ook at any illustrations or diagrams.
  They may provide you with additional information to answer the question.
- ☐ **U**nderstand the way you are asked to answer the question.
  - ☐ Graph
  - ☐ Chart
  - ☐ Diagram
  - ☐ Complete sentences
  - ☐ Phrase
  - ☐ Filled circle
- ☐ **E**xamine the information given.
  - ☐ Reread the questions.
  - ☐ Underline key words or phrases.
  - ☐ Think about what the questions are asking.
- ☐ **S**ee if your answers match the questions.

98 COPYING IS PROHIBITED © Englefield & Associates, Inc.

Note: Student answers may vary. Example responses are for use as a guide.

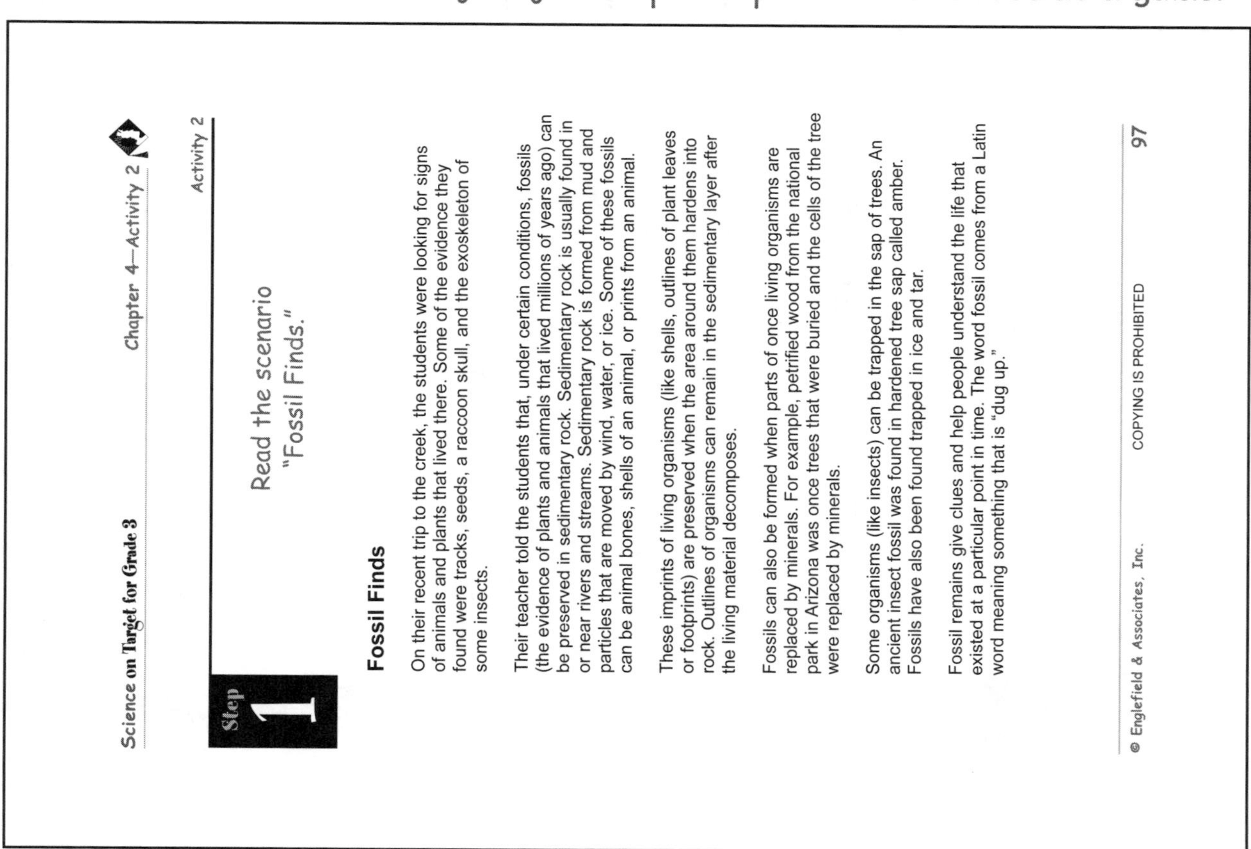

Science on Target for Grade 3 Chapter 4—Activity 2

Activity 2

Step 1 Read the scenario "Fossil Finds."

Fossil Finds

On their recent trip to the creek, the students were looking for signs of animals and plants that lived there. Some of the evidence they found were tracks, seeds, a raccoon skull, and the exoskeleton of some insects.

Their teacher told the students that, under certain conditions, fossils (the evidence of plants and animals that lived millions of years ago) can be preserved in sedimentary rock. Sedimentary rock is usually found in or near rivers and streams. Sedimentary rock is formed from mud and particles that are moved by wind, water, or ice. Some of these fossils can be animal bones, shells of an animal, or prints from an animal.

These imprints of living organisms (like shells, outlines of plant leaves or footprints) are preserved when the area around them hardens into rock. Outlines of organisms can remain in the sedimentary layer after the living material decomposes.

Fossils can also be formed when parts of once living organisms are replaced by minerals. For example, petrified wood from the national park in Arizona was once trees that were buried and the cells of the tree were replaced by minerals.

Some organisms (like insects) can be trapped in the sap of trees. An ancient insect fossil was found in hardened tree sap called amber. Fossils have also been found trapped in ice and tar.

Fossil remains give clues and help people understand the life that existed at a particular point in time. The word fossil comes from a Latin word meaning something that is "dug up."

© Englefield & Associates, Inc. COPYING IS PROHIBITED 97

Activity 2

Step 3

Use the information from "Fossil Finds" to complete the graphic organizer.

Use words and phrases from the scenario "Fossil Finds" to describe how the specific kind of fossils listed below would have been preserved.

Fossil	How it was preserved
mammoth bone	The bone might have been trapped in the tar of the LaBrea tar pits.
imprint of fern	Sedimentary rock formed around the fern.
bee found in amber	The bee is trapped in hardened tree sap called amber.
bird footprints	Sedimentary rock formed over the bird footprints.
petrified wood	The cells of the wood were replaced by minerals.

Note: Student answers may vary. Example responses are for use as a guide.

Activity 2

Step 4

Answer the following questions for "Fossil Finds" using information from your graphic organizer.

1. The word fossil comes from the Latin word meaning "dug up." Why is this a good word to describe evidence of organisms that lived long ago?

   Some of the fossils are found deep in the layers of the earth. So scientists must use tools to get (dig) them from the area.

2. There is a famous fossil area in California called the La Brea Tar Pits. Tar is a thick gooey material. What kind of fossils do you think are most often found there?

   bones and skeletons of animals

   Why?

   The gooey material surrounds the bones to preserve them.

Step 2

Complete the Checklist "Clues for Success."

The checklist will help you to read and think like a scientist.

Clues for Success

- ☐ **C**arefully read the information.
- ☐ **L**ook at any illustrations or diagrams.
  They may provide you with additional information to answer the question.
- ☐ **U**nderstand the way you are asked to answer the question.
  - ☐ Graph
  - ☐ Chart
  - ☐ Diagram
  - ☐ Complete sentences
  - ☐ Phrase
  - ☐ Filled circle
- ☐ **E**xamine the information given.
  - ☐ Reread the questions.
  - ☐ Underline key words or phrases.
  - ☐ Think about what the questions are asking.
- ☐ **S**ee if your answers match the questions.

Note: Student answers may vary. Example responses are for use as a guide.

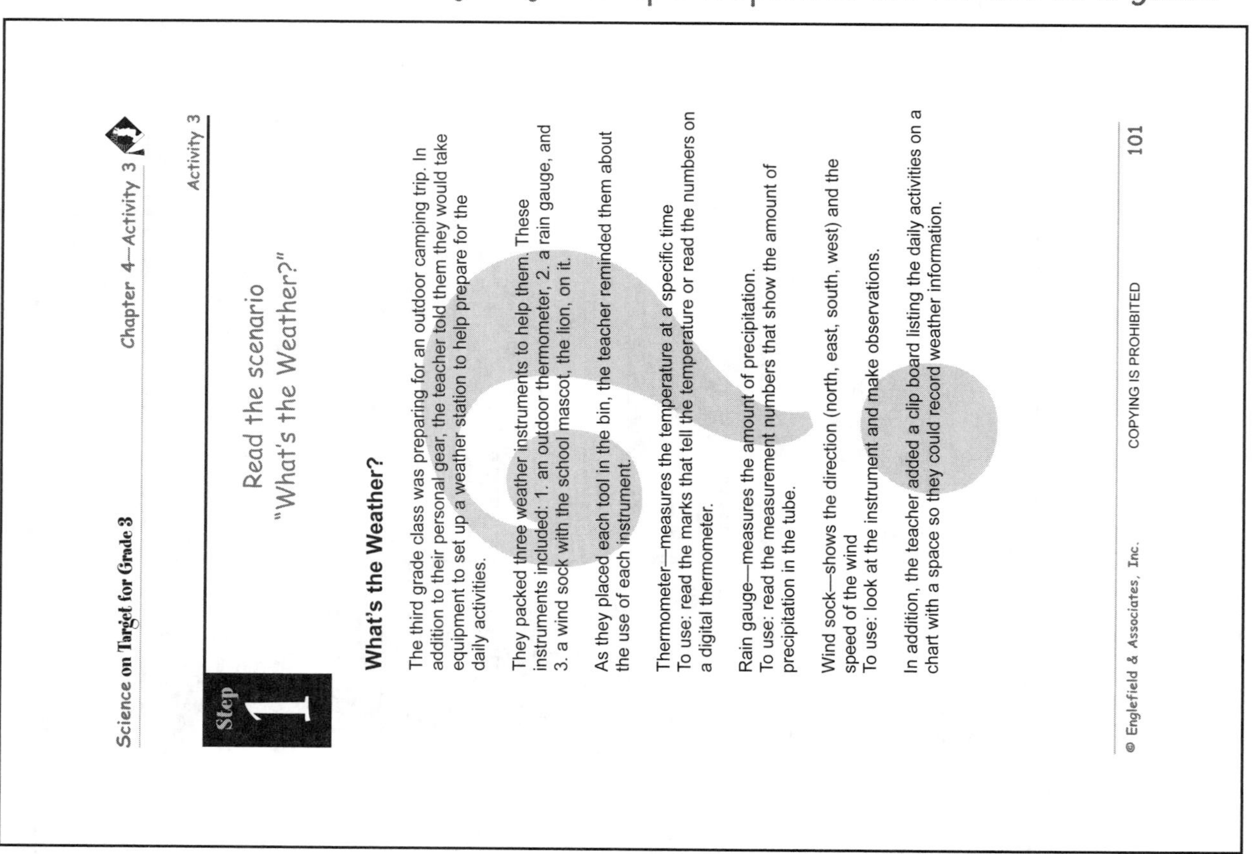

Step 1

Read the scenario
"What's the Weather?"

What's the Weather?

The third grade class was preparing for an outdoor camping trip. In addition to their personal gear, the teacher told them they would take equipment to set up a weather station to help prepare for the daily activities.

They packed three weather instruments to help them. These instruments included: 1. an outdoor thermometer, 2. a rain gauge, and 3. a wind sock with the school mascot, the lion, on it.

As they placed each tool in the bin, the teacher reminded them about the use of each instrument.

Thermometer—measures the temperature at a specific time
To use: read the marks that tell the temperature or read the numbers on a digital thermometer.

Rain gauge—measures the amount of precipitation.
To use: read the measurement numbers that show the amount of precipitation in the tube.

Wind sock—shows the direction (north, east, south, west) and the speed of the wind
To use: look at the instrument and make observations.

In addition, the teacher added a clip board listing the daily activities on a chart with a space so they could record weather information.

Activity 3

Step 4

Answer the following questions for "What's the Weather?" using information from your graphic organizer.

Answer the questions with a completely filled circle or circles. Then explain your answer with a complete sentence.

1. What season or seasons might the rain gauge have water in it?

● Spring ○ Summer ○ Fall ○ Winter

Explain your answer

Rain often falls in the spring.

2. During which season is the rain gauge most likely to have snow in it?

○ Spring ○ Summer ○ Fall ● Winter

Explain your answer

In winter, the temperature is low. So, the water in the air freezes turning it to snow.

Note: Student answers may vary. Example responses are for use as a guide.

Activity 3

Step 3

Use the information from "What's the Weather?" to complete the graphic organizer.

Use the information about the weather tools in the scenario, "What's the Weather?" to select the tool that will provide the needed information. Use a completely filled circle to mark your answer choice.

Day 1	thermometer	rain gauge	wind sock
Did the outdoor temperature reach the 68 degrees needed for the pool to open?	●	○	○
The direction they should fly their kites to avoid the trees	○	○	●
If it rained at camp after they arrived	○	●	○

Day 2	thermometer	rain gauge	wind sock
How much rain fell during the night?	○	●	○
If wind was blowing too hard for the paper airplane contest	○	○	●
If they should take a sweatshirt on the hike	●	○	○

3. During which season or seasons might the thermometer read 72 degrees?

○ Spring ● Summer ○ Fall ○ Winter

Explain your answer

A temperature of 72° is usually recorded during the summer in most areas.

4. During which season or seasons would you expect the wind sock to move most briskly?

○ Spring ○ Summer ● Fall ○ Winter

Explain your answer

The wind is brisk in the fall. The leaves often blow off the trees.

Note: Student answers may vary. Example responses are for use as a guide.

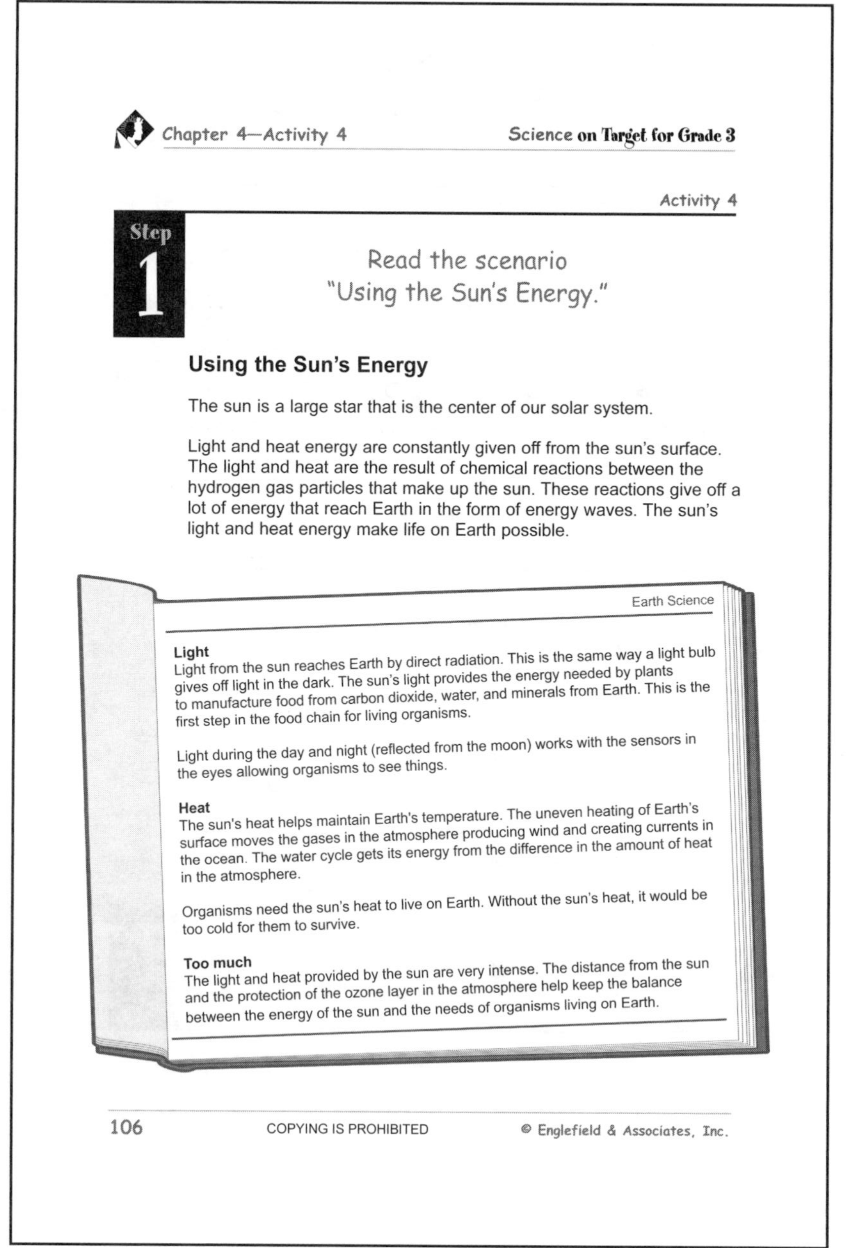

Activity 4

Step 1

Read the scenario "Using the Sun's Energy."

Using the Sun's Energy

The sun is a large star that is the center of our solar system.

Light and heat energy are constantly given off from the sun's surface. The light and heat are the result of chemical reactions between the hydrogen gas particles that make up the sun. These reactions give off a lot of energy that reach Earth in the form of energy waves. The sun's light and heat energy make life on Earth possible.

Earth Science

Light
Light from the sun reaches Earth by direct radiation. This is the same way a light bulb gives off light in the dark. The sun's light provides the energy needed by plants to manufacture food from carbon dioxide, water, and minerals from Earth. This is the first step in the food chain for living organisms.

Light during the day and night (reflected from the moon) works with the sensors in the eyes allowing organisms to see things.

Heat
The sun's heat helps maintain Earth's temperature. The uneven heating of Earth's surface moves the gases in the atmosphere producing wind and creating currents in the ocean. The water cycle gets its energy from the difference in the amount of heat in the atmosphere.

Organisms need the sun's heat to live on Earth. Without the sun's heat, it would be too cold for them to survive.

Too much
The light and heat provided by the sun are very intense. The distance from the sun and the protection of the ozone layer in the atmosphere help keep the balance between the energy of the sun and the needs of organisms living on Earth.

Step 3

Use the information from "Using the Sun's Energy" to complete the graphic organizer.

Activity 4

The Sun's Energy

I. The sun's energy is created by chemical reactions

II. Two forms of sun's energy are
- A. light
- B. heat

III. Uses for light energy
- A. energy for plants to make food
- B. some organisms use light to see

IV. Uses for heat energy
- A. maintain Earth's temperature
- B. wind
- C. ocean currents
- D. operate water cycles

V. Protection from the intense light and heat of the sun is provided by
- A. distance from Earth
- B. ozone layer in atmosphere

Note: Student answers may vary. Example responses are for use as a guide.

Step 2

Complete the Checklist "Clues for Success."

The checklist will help you to read and think like a scientist.

Activity 4

Clues for Success

☐ **C**arefully read the information.

☐ **L**ook at any illustrations or diagrams.
They may provide you with additional information to answer the question.

☐ **U**nderstand the way you are asked to answer the question.
- ☐ Graph
- ☐ Chart
- ☐ Diagram
- ☐ Complete sentences
- ☐ Phrase
- ☐ Filled circle

☐ **E**xamine the information given.
- ☐ Reread the questions.
- ☐ Underline key words or phrases.
- ☐ Think about what the questions are asking.

☐ **S**ee if your answers match the questions.

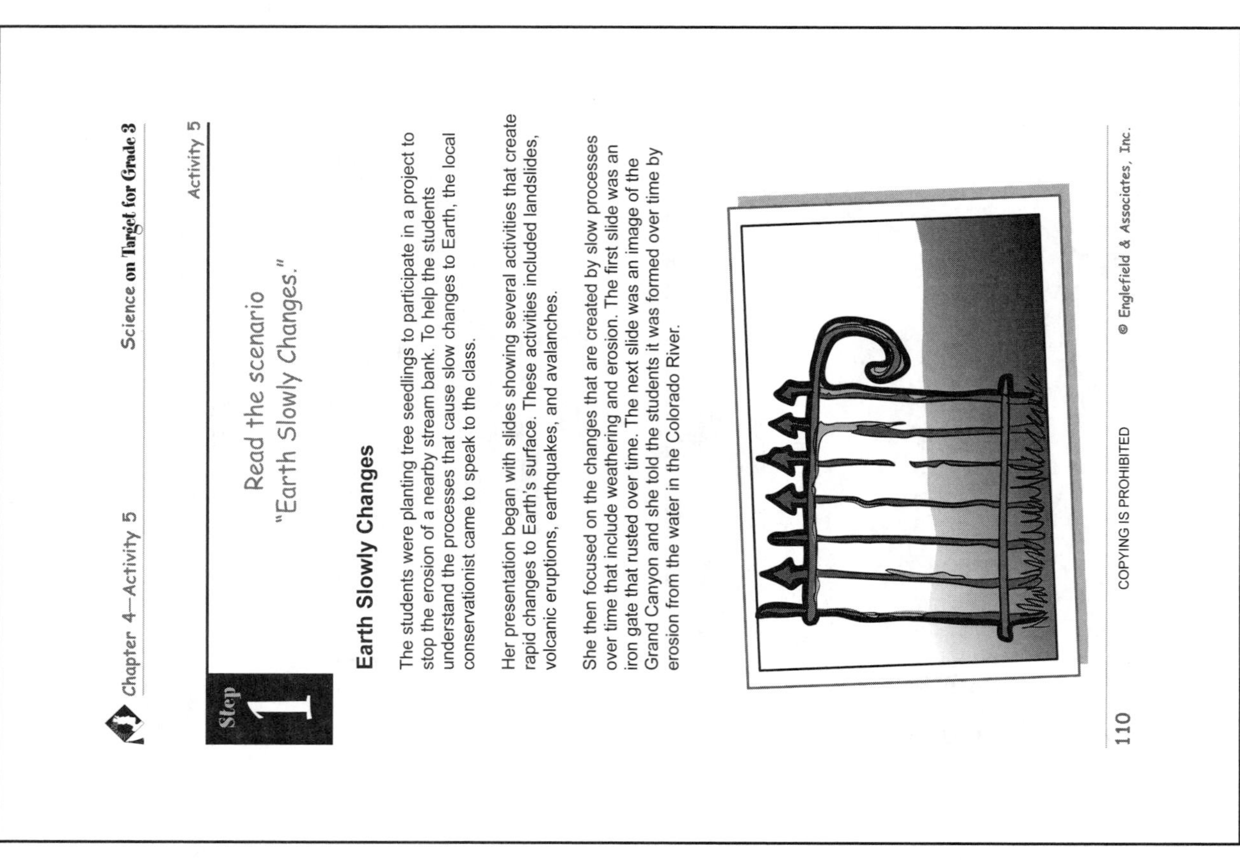

Step 1

Read the scenario "Earth Slowly Changes."

Activity 5

Earth Slowly Changes

The students were planting tree seedlings to participate in a project to stop the erosion of a nearby stream bank. To help the students understand the processes that cause slow changes to Earth, the local conservationist came to speak to the class.

Her presentation began with slides showing several activities that create rapid changes to Earth's surface. These activities included landslides, volcanic eruptions, earthquakes, and avalanches.

She then focused on the changes that are created by slow processes over time that include weathering and erosion. The first slide was an iron gate that rusted over time. The next slide was an image of the Grand Canyon and she told the students it was formed over time by erosion from the water in the Colorado River.

Note: Student answers may vary. Example responses are for use as a guide.

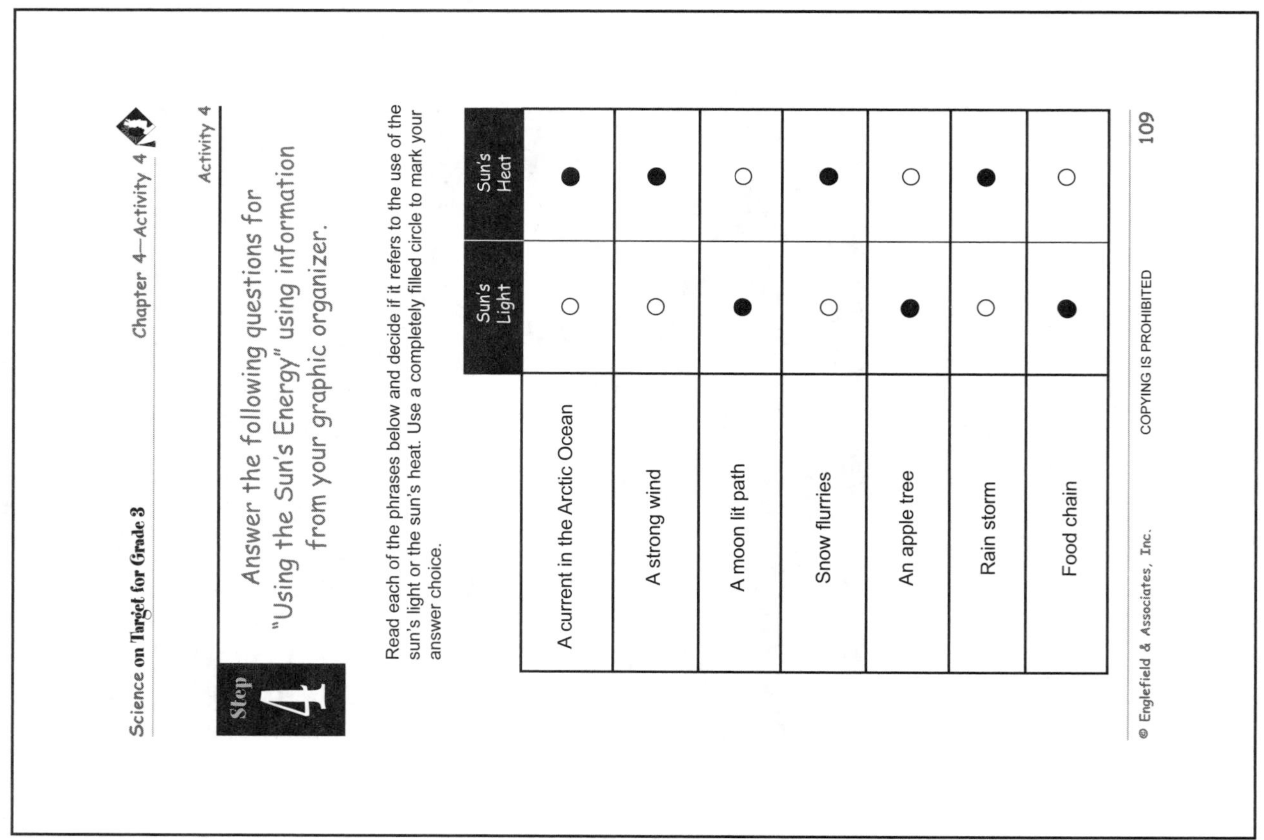

Step 4

Answer the following questions for "Using the Sun's Energy" using information from your graphic organizer.

Activity 4

Read each of the phrases below and decide if it refers to the use of the sun's light or the sun's heat. Use a completely filled circle to mark your answer choice.

	Sun's Light	Sun's Heat
A current in the Arctic Ocean	○	●
A strong wind	○	●
A moon lit path	●	○
Snow flurries	○	●
An apple tree	●	○
Rain storm	○	●
Food chain	●	○

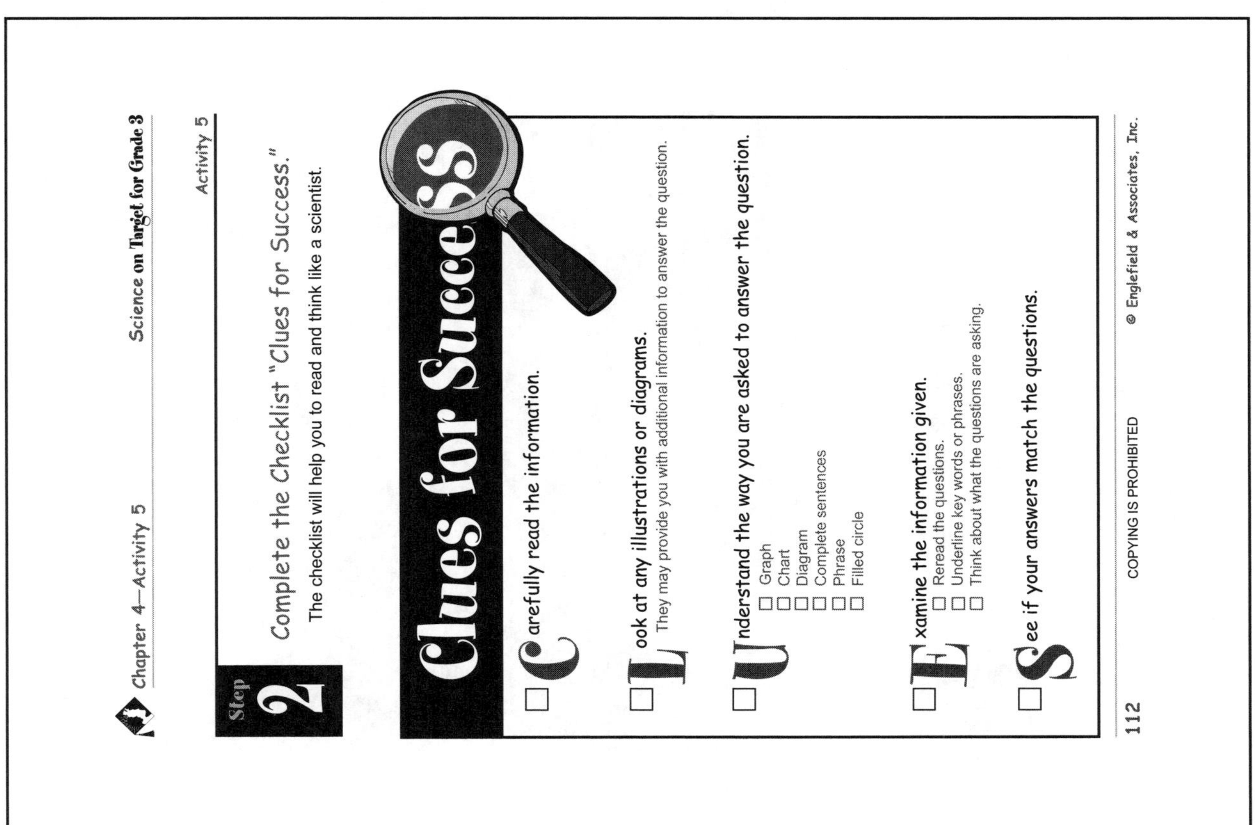

Activity 5

Step 2

Complete the Checklist "Clues for Success."

The checklist will help you to read and think like a scientist.

Clues for Success

- ☐ **C**arefully read the information.
- ☐ **L**ook at any illustrations or diagrams.
  They may provide you with additional information to answer the question.
- ☐ **U**nderstand the way you are asked to answer the question.
  - ☐ Graph
  - ☐ Chart
  - ☐ Diagram
  - ☐ Complete sentences
  - ☐ Phrase
  - ☐ Filled circle
- ☐ **E**xamine the information given.
  - ☐ Reread the questions.
  - ☐ Underline key words or phrases.
  - ☐ Think about what the questions are asking.
- ☐ **S**ee if your answers match the questions.

Note: Student answers may vary. Example responses are for use as a guide.

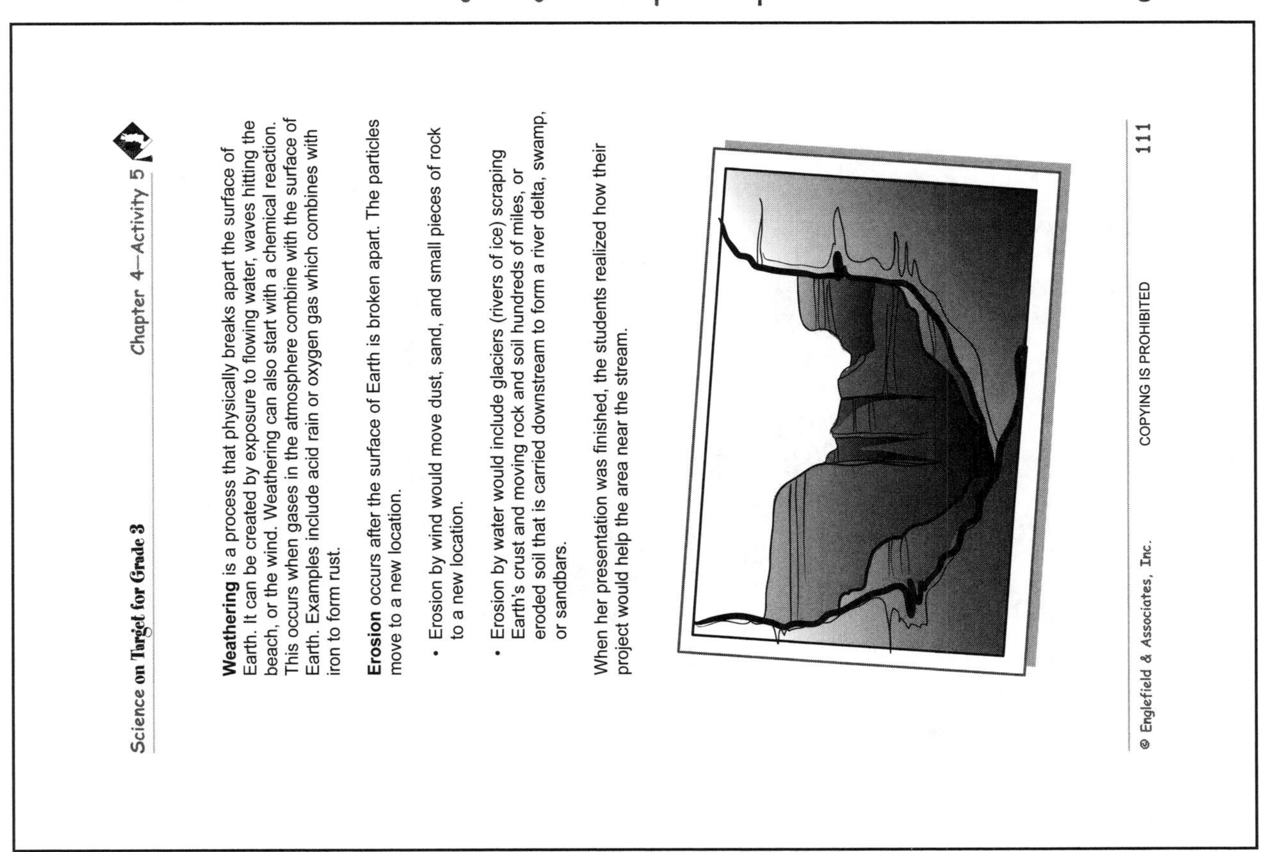

Weathering is a process that physically breaks apart the surface of Earth. It can be created by exposure to flowing water, waves hitting the beach, or the wind. Weathering can also start with a chemical reaction. This occurs when gases in the atmosphere combine with the surface of Earth. Examples include acid rain or oxygen gas which combines with iron to form rust.

Erosion occurs after the surface of Earth is broken apart. The particles move to a new location.

- Erosion by wind would move dust, sand, and small pieces of rock to a new location.
- Erosion by water would include glaciers (rivers of ice) scraping Earth's crust and moving rock and soil hundreds of miles, or eroded soil that is carried downstream to form a river delta, swamp, or sandbars.

When her presentation was finished, the students realized how their project would help the area near the stream.

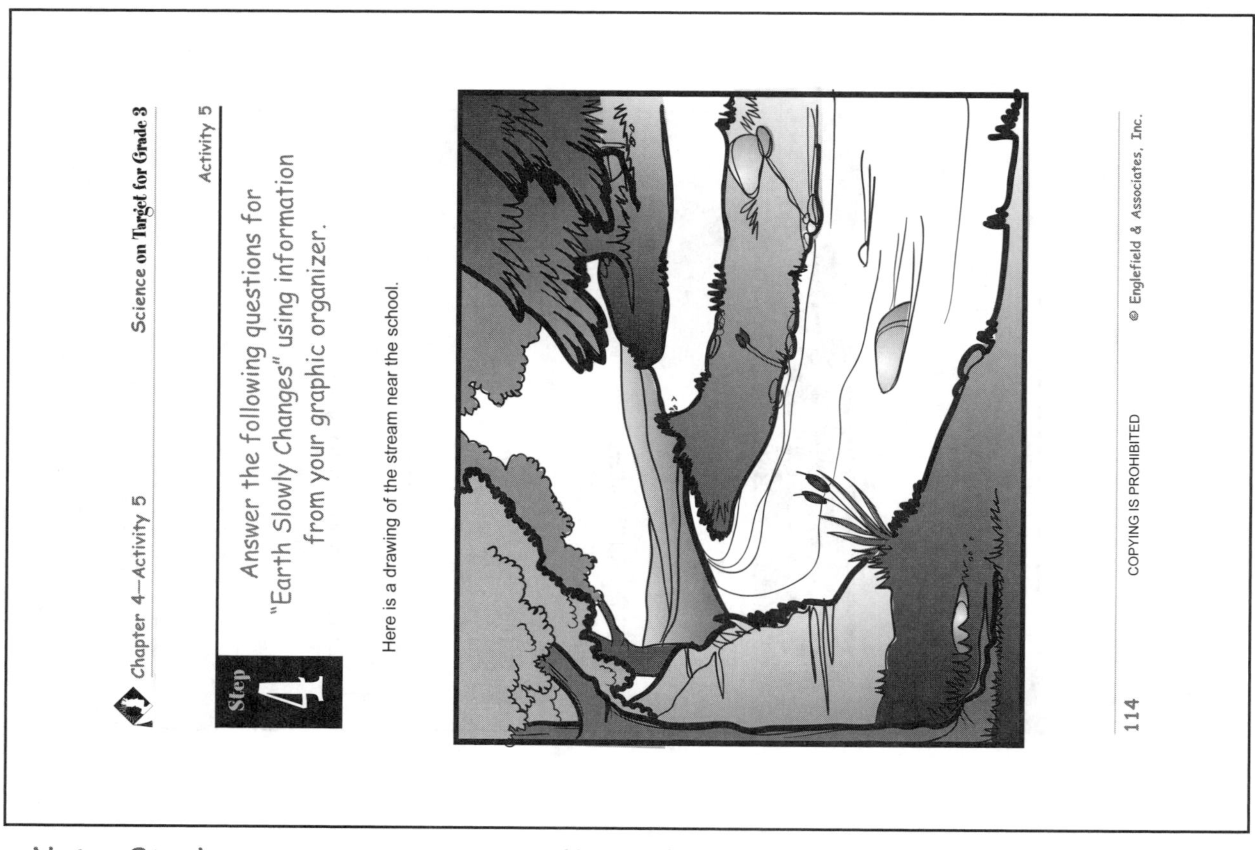

Step 4

Answer the following questions for "Earth Slowly Changes" using information from your graphic organizer.

Here is a drawing of the stream near the school.

Note: Student answers may vary. Example responses are for use as a guide.

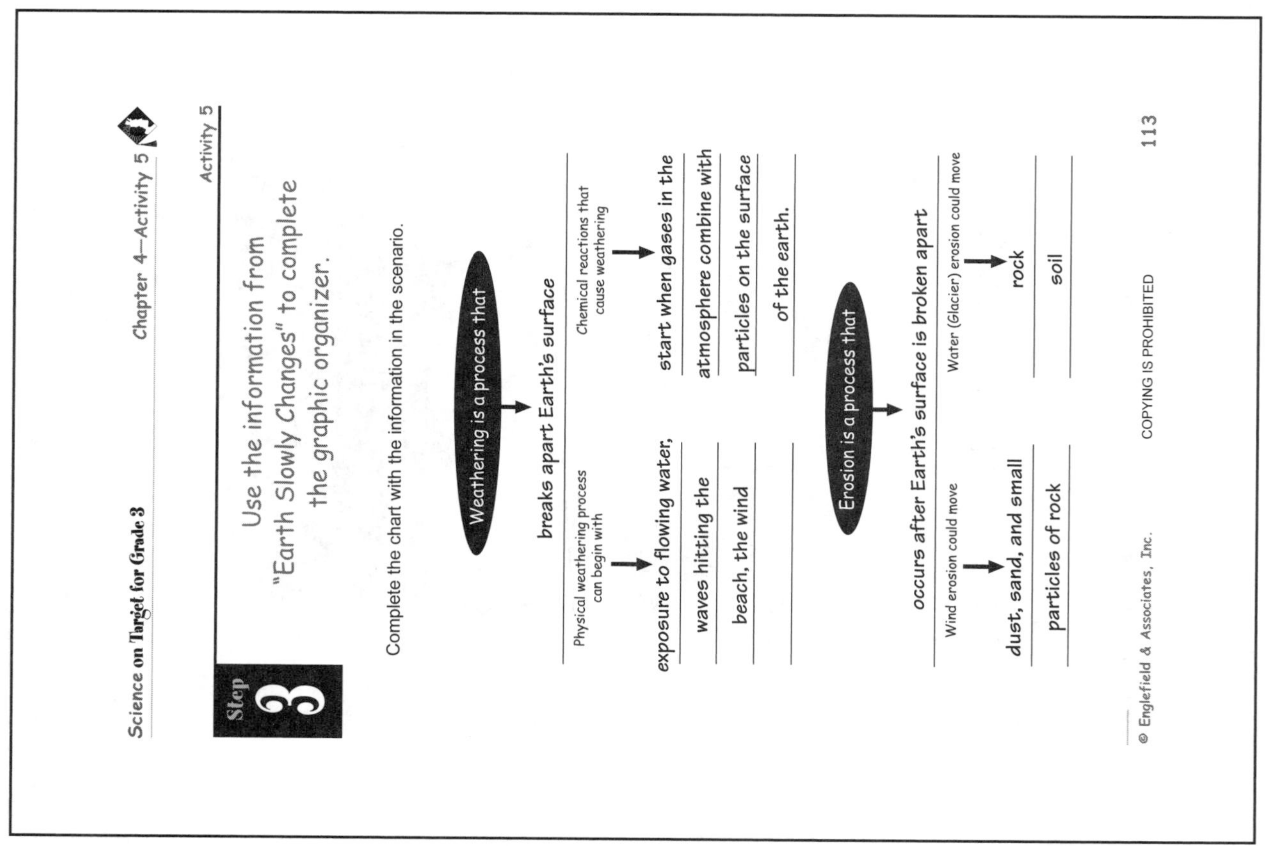

Step 3

Use the information from "Earth Slowly Changes" to complete the graphic organizer.

Complete the chart with the information in the scenario.

Weathering is a process that → breaks apart Earth's surface

Physical weathering process can begin with → exposure to flowing water, waves hitting the beach, the wind

Chemical reactions that cause weathering → start when gases in the atmosphere combine with particles on the surface of the earth.

Erosion is a process that → occurs after Earth's surface is broken apart

Wind erosion could move → dust, sand, and small particles of rock

Water (Glacier) erosion could move → rock soil

2. Draw an illustration of how the stream bank might look in 20 years with the help from seedlings planted to help stop weathering and erosion. Then, use a complete sentence to explain your drawing.

There would be more vegetation to stop erosion of the land.

Note: Student answers may vary. Example responses are for use as a guide.

1. Draw an illustration of how the stream bank might look with the effects of weathering and erosion in 20 years. Then, use a complete sentence to explain your drawing.

The roots of the trees will be exposed and the river would cut away into the land. More rocks will be exposed.

Chapter 5

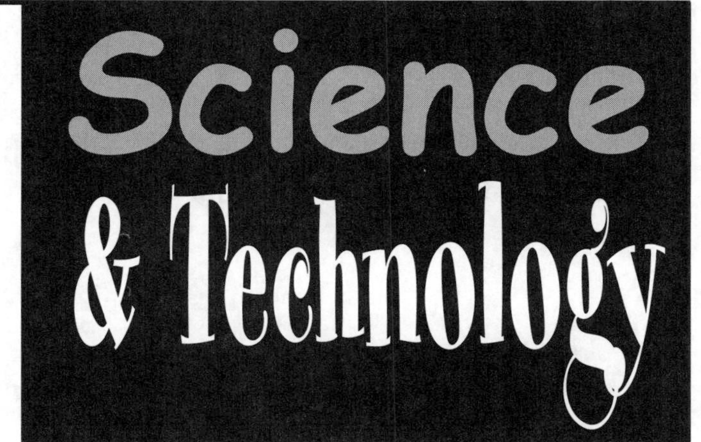

Chapter 5 of the *Science on Target for Grade 3 Student Workbook*, covers the National Content Standards for Science & Technology. The standards are as follows:

Science & Technology

Abilities of technological designs

- Identify a simple problem.
- Propose a solution.
- Implement a proposed solution.
- Evaluate proposed solutions.
- Evaluate a product or design.
- Communicate a problem, design, and solution.

Understanding about science and technology

- People have questions about their world. Science is a way of answering questions and explaining the natural world.
- People invented tools and techniques to solve problems.
- Scientists and engineers often work in teams with different individuals doing different things that contribute to results.
- Women and men of all ages and backgrounds engage in scientific and technological work.
- Tools help scientists make better observations, measurements, and equipment for investigations. (See, measure, and do things they could not otherwise do.)

Abilities to distinguish between natural objects and objects made by humans

- Some objects are designed in nature and others have been designed by people.
- Objects can be categorized into two groups, natural and designed.

All of the pages from Chapter 5 of the *Science on Target for Grade 3 Student Workbook*, are reproduced in this Parent/Teacher Edition in reduced-page format with sample answers. These activities will help your students develop the skills necessary to understand Science and Technology.

Students should use the "Clues for Success" Checklist, for each activity in this section, as a tool to help them do their best work.

Step 1 Activity 1

Read the scenario
"The Earmuff Capital of the World."

The Earmuff Capital of the World

The advertisement in the paper said:

The article on the next page told the story of Chester Greenwood.

Note: Student answers may vary. Example responses are for use as a guide.

Chapter 5

The activities in this section of the book will focus on Science and Technology.

These activities will help you develop abilities to:

- communicate, design and propose a solution to a question,
- understand that many different kinds of people do Science and create technology to learn about the world and make life easier, and
- know the difference between natural and man-made objects.

Use the "Clues for Success" Checklist, as you complete each activity in this section, as a tool to help you do your best work.

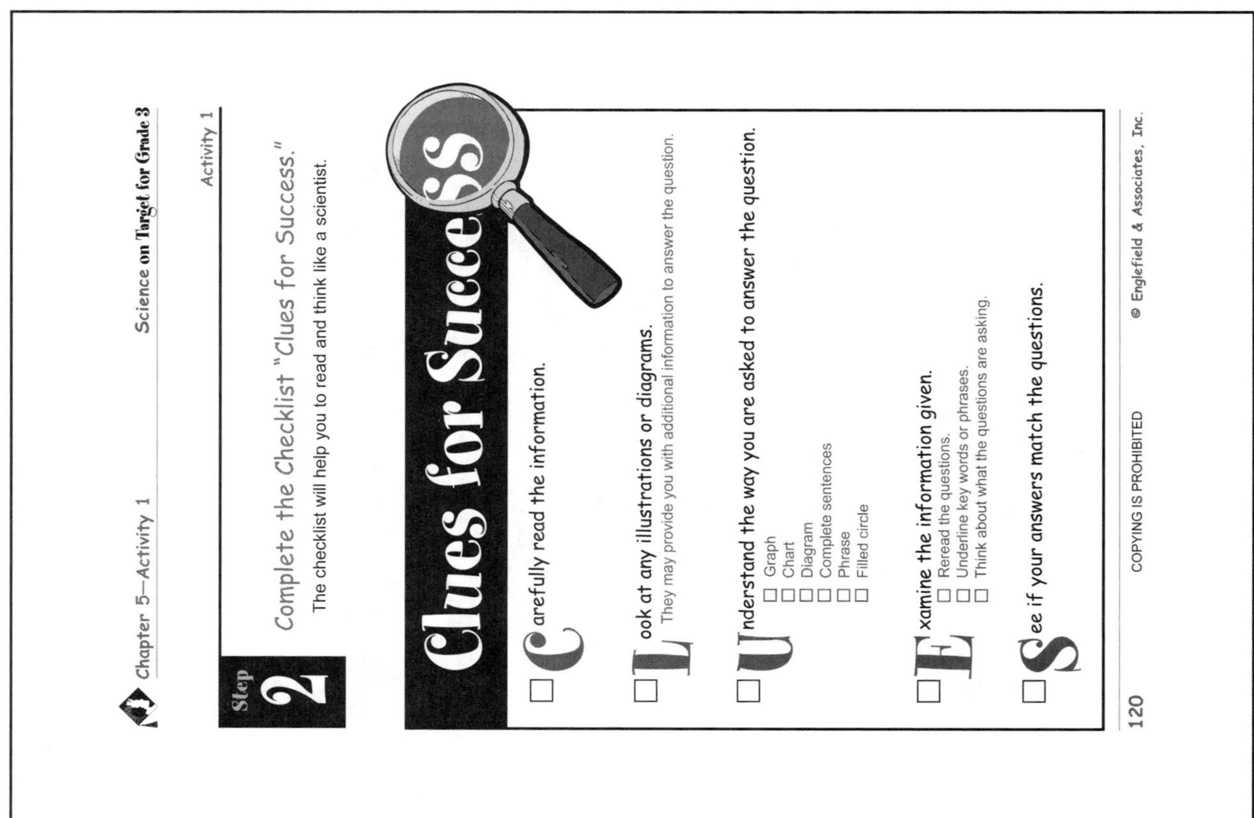

Chapter 5—Activity 1 Science on Target for Grade 3

Activity 1

Step 2 Complete the Checklist "Clues for Success."

The checklist will help you to read and think like a scientist.

Clues for Success

- ☐ **C**arefully read the information.
- ☐ **L**ook at any illustrations or diagrams.
  They may provide you with additional information to answer the question.
- ☐ **U**nderstand the way you are asked to answer the question.
  - ☐ Graph
  - ☐ Chart
  - ☐ Diagram
  - ☐ Complete sentences
  - ☐ Phrase
  - ☐ Filled circle
- ☐ **E**xamine the information given.
  - ☐ Reread the questions.
  - ☐ Underline key words or phrases.
  - ☐ Think about what the questions are asking.
- ☐ **S**ee if your answers match the questions.

120 COPYING IS PROHIBITED © Englefield & Associates, Inc.

Note: Student answers may vary. Example responses are for use as a guide.

Science on Target for Grade 3 Chapter 5—Activity 1

CHESTER GREENWOOD DAY

Farmington, Maine, will be celebrating "Chester Greenwood Day" to honor a young man who made a major contribution to cold weather protection when he was 15 years old. Chester Greenwood was born in Farmington, Maine, in 1858. Chester loved to ice skate but while he was skating his ears got cold. When he wrapped a scarf around his ears, his ears itched. The scarf flapped as he skated and sometimes even covered his eyes.

To solve his problem, Chester formed two wire loops (big enough to cover his ears) and asked his grandmother to cover them with fur. He used these ear covers when he skated.

He improved the ear covers by adding a headband to keep them in place. He named his invention, Greenwood's Champion Ear Protectors. He set up a factory called Greenwood's Ear Protection Factory. Chester even sold his ear protectors to help U.S. soldiers in World War I protect their ears during battle.

In 1977, Maine's legislature declared December 21 "Chester Greenwood Day" to celebrate his invention of this cold weather protection gear.

© Englefield & Associates, Inc. COPYING IS PROHIBITED 119

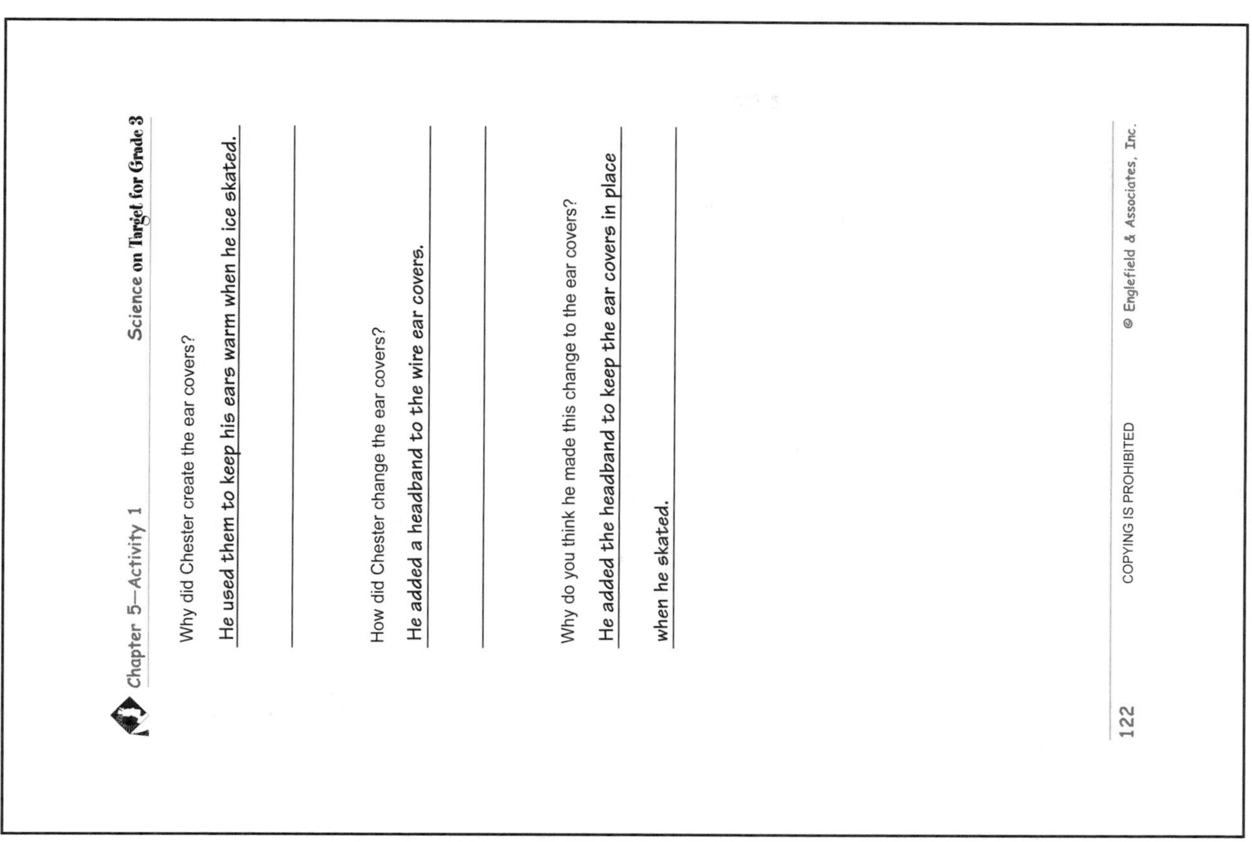

Why did Chester create the ear covers?

He used them to keep his ears warm when he ice skated.

How did Chester change the ear covers?

He added a headband to the wire ear covers.

Why do you think he made this change to the ear covers?

He added the headband to keep the ear covers in place when he skated.

Note: Student answers may vary. Example responses are for use as a guide.

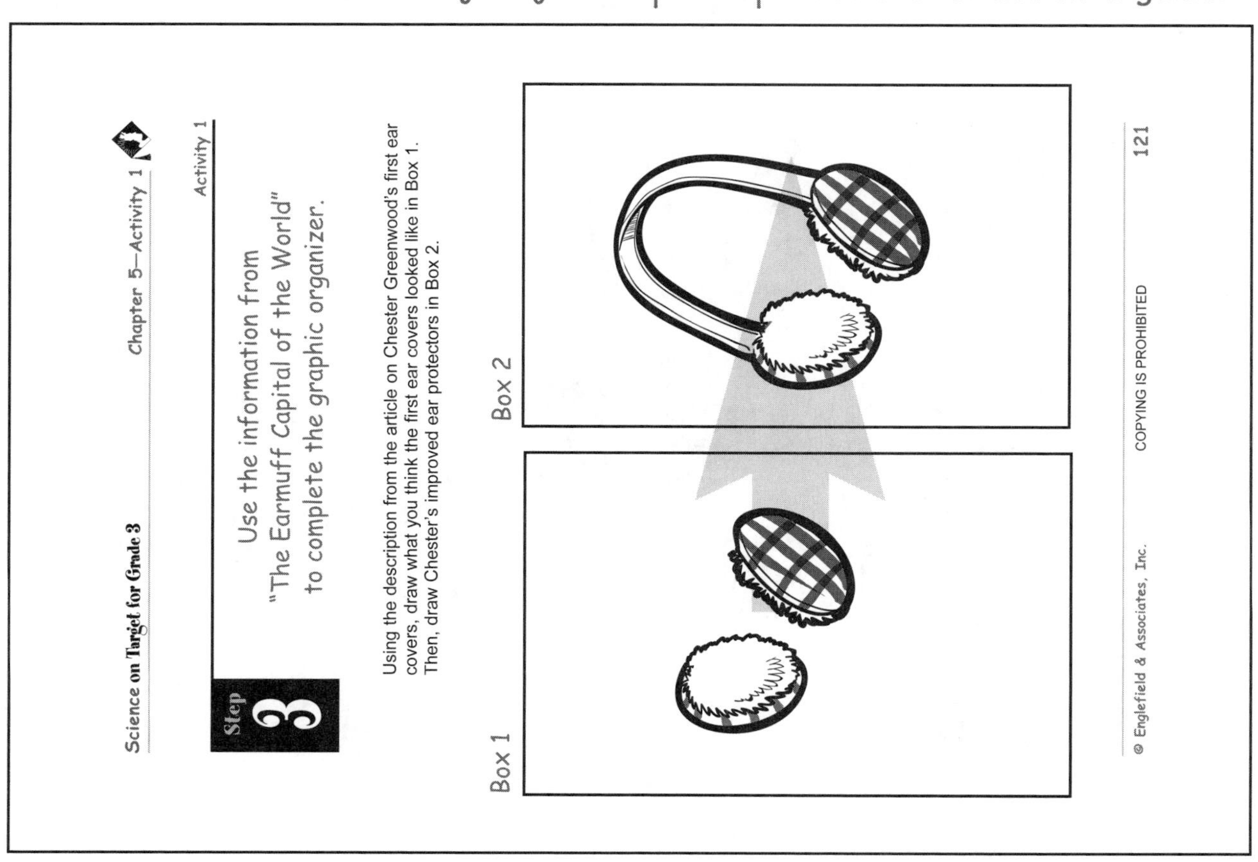

Step 3 Activity 1

Use the information from "The Earmuff Capital of the World" to complete the graphic organizer.

Using the description from the article on Chester Greenwood's first ear covers, draw what you think the first ear covers looked like in Box 1. Then, draw Chester's improved ear protectors in Box 2.

Box 1

Box 2

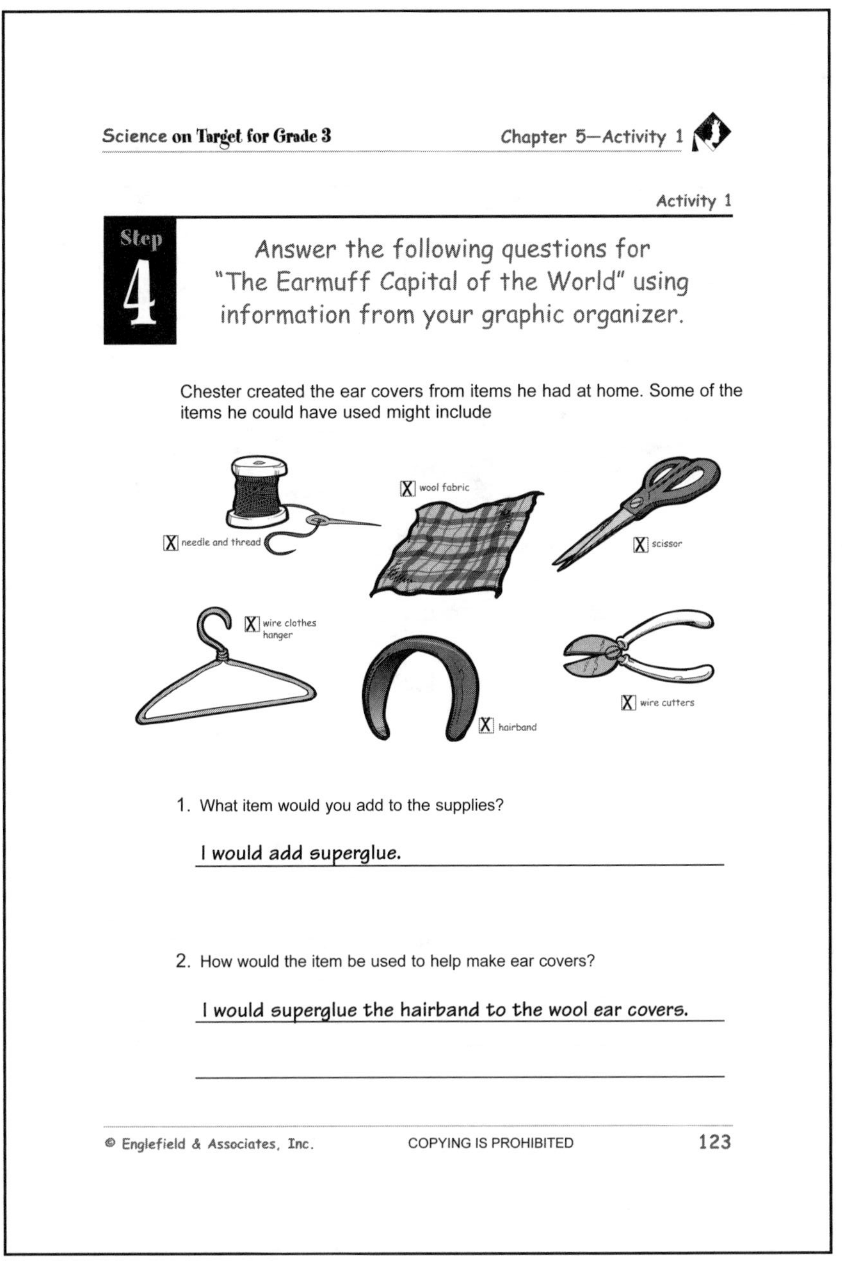
Science on Target for Grade 3 Chapter 5—Activity 1

Activity 1

Step 4

Answer the following questions for "The Earmuff Capital of the World" using information from your graphic organizer.

Chester created the ear covers from items he had at home. Some of the items he could have used might include

1. What item would you add to the supplies?

I would add superglue.

2. How would the item be used to help make ear covers?

I would superglue the hairband to the wool ear covers.

© Englefield & Associates, Inc. COPYING IS PROHIBITED 123

Note: Student answers may vary. Example responses are for use as a guide.

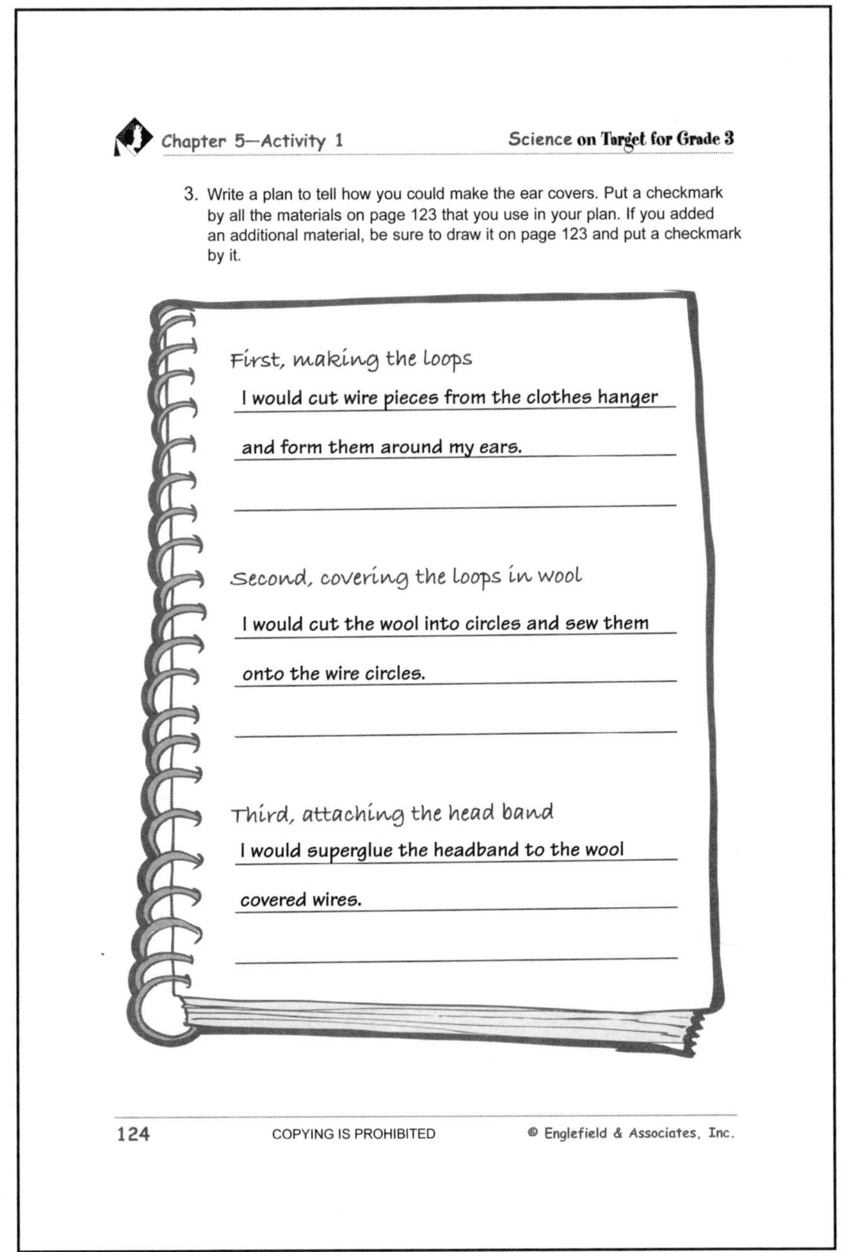
Chapter 5—Activity 1 Science on Target for Grade 3

3. Write a plan to tell how you could make the ear covers. Put a checkmark by all the materials on page 123 that you use in your plan. If you added an additional material, be sure to draw it on page 123 and put a checkmark by it.

First, making the loops

I would cut wire pieces from the clothes hanger and form them around my ears.

Second, covering the loops in wool

I would cut the wool into circles and sew them onto the wire circles.

Third, attaching the head band

I would superglue the headband to the wool covered wires.

124 COPYING IS PROHIBITED © Englefield & Associates, Inc.

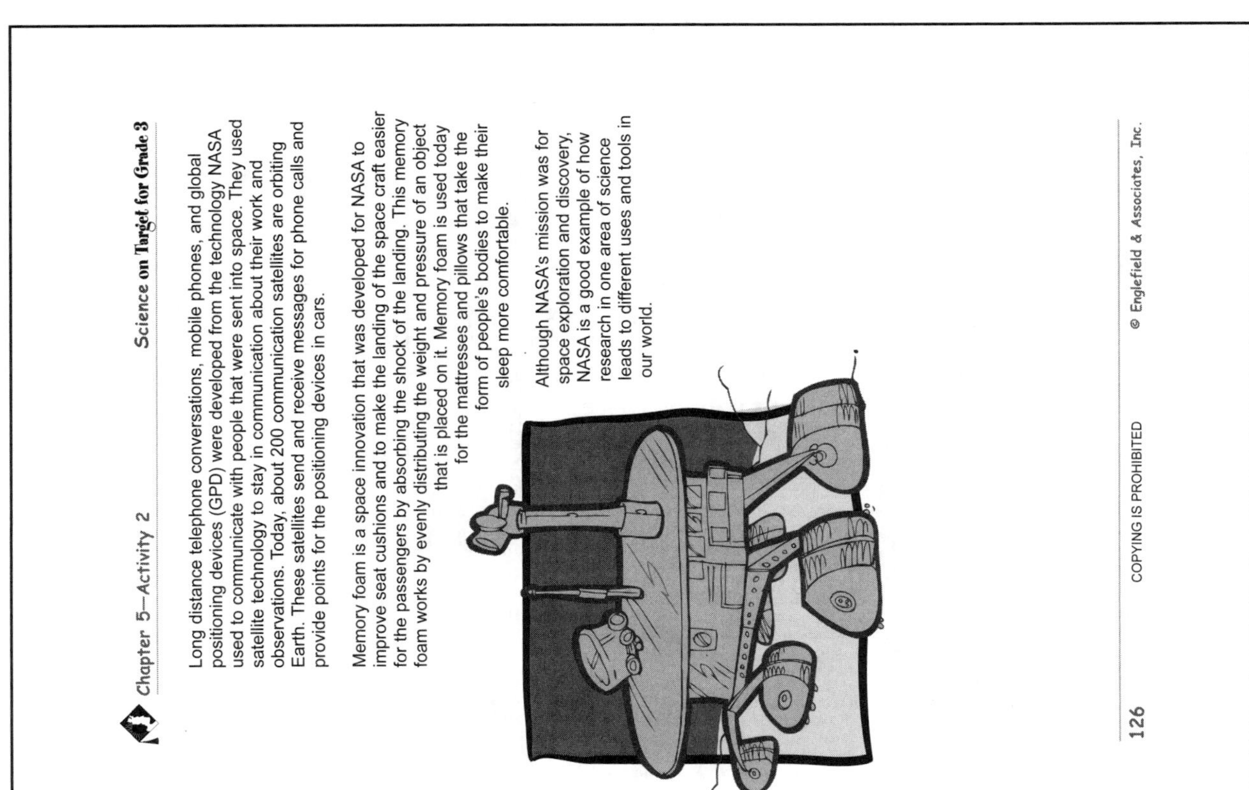

Chapter 5—Activity 2 Science on Target for Grade 3

Long distance telephone conversations, mobile phones, and global positioning devices (GPD) were developed from the technology NASA used to communicate with people that were sent into space. They used satellite technology to stay in communication about their work and observations. Today, about 200 communication satellites are orbiting Earth. These satellites send and receive messages for phone calls and provide points for the positioning devices in cars.

Memory foam is a space innovation that was developed for NASA to improve seat cushions and to make the landing of the space craft easier for the passengers by absorbing the shock of the landing. This memory foam works by evenly distributing the weight and pressure of an object that is placed on it. Memory foam is used today for the mattresses and pillows that take the form of people's bodies to make their sleep more comfortable.

Although NASA's mission was for space exploration and discovery, NASA is a good example of how research in one area of science leads to different uses and tools in our world.

126 COPYING IS PROHIBITED © Englefield & Associates, Inc.

Note: Student answers may vary. Example responses are for use as a guide.

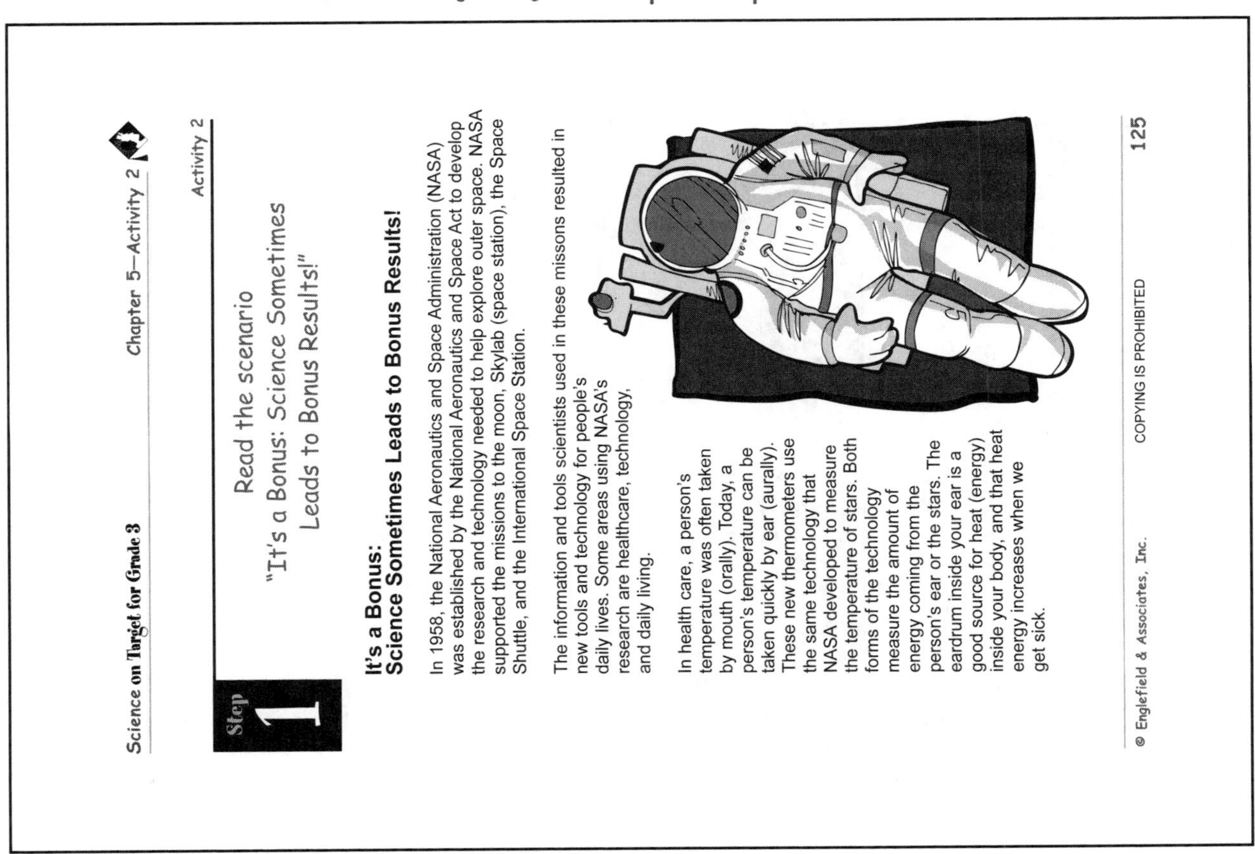

Science on Target for Grade 3 Chapter 5—Activity 2

Step 1

Activity 2

Read the scenario "It's a Bonus: Science Sometimes Leads to Bonus Results!"

It's a Bonus:
Science Sometimes Leads to Bonus Results!

In 1958, the National Aeronautics and Space Administration (NASA) was established by the National Aeronautics and Space Act to develop the research and technology needed to help explore outer space. NASA supported the missions to the moon, Skylab (space station), the Space Shuttle, and the International Space Station.

The information and tools scientists used in these missons resulted in new tools and technology for people's daily lives. Some areas using NASA's research are healthcare, technology, and daily living.

In health care, a person's temperature was often taken by mouth (orally). Today, a person's temperature can be taken quickly by ear (aurally). These new thermometers use the same technology that NASA developed to measure the temperature of stars. Both forms of the technology measure the amount of energy coming from the person's ear or the stars. The eardrum inside your ear is a good source for heat (energy) inside your body, and that heat energy increases when we get sick.

© Englefield & Associates, Inc. COPYING IS PROHIBITED 125

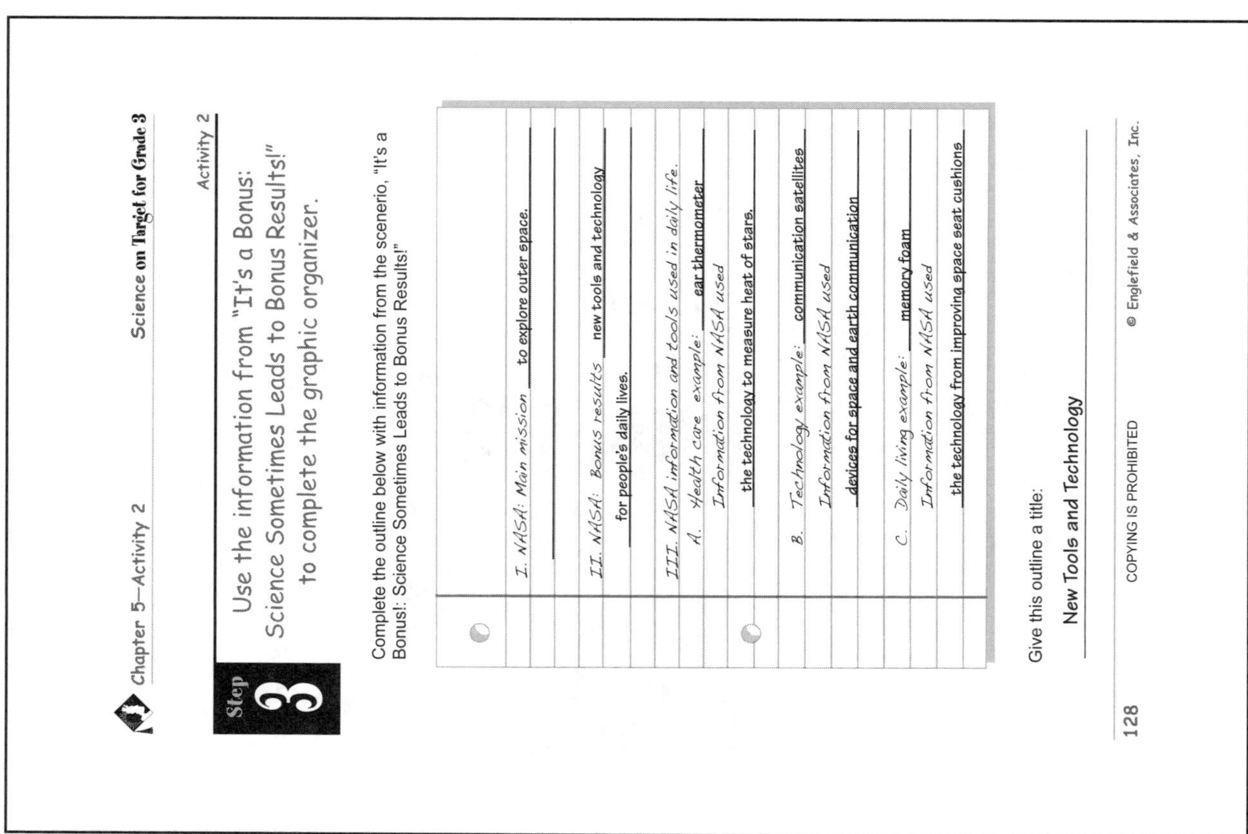
Chapter 5—Activity 2 Science on Target for Grade 3

Activity 2

Step 3

Use the information from "It's a Bonus: Science Sometimes Leads to Bonus Results!" to complete the graphic organizer.

Complete the outline below with information from the scenerio, "It's a Bonus!: Science Sometimes Leads to Bonus Results!"

I. NASA: Main mission to explore outer space.

II. NASA: Bonus results new tools and technology for people's daily lives.

III. NASA information and tools used in daily life.

A. Health care example: ear thermometer

Information from NASA used the technology to measure heat of stars.

B. Technology example: communication satellites

Information from NASA used devices for space and earth communication

C. Daily living example: memory foam

Information from NASA used the technology from improving space seat cushions

Give this outline a title:

New Tools and Technology

128 COPYING IS PROHIBITED © Englefield & Associates, Inc.

Note: Student answers may vary. Example responses are for use as a guide.

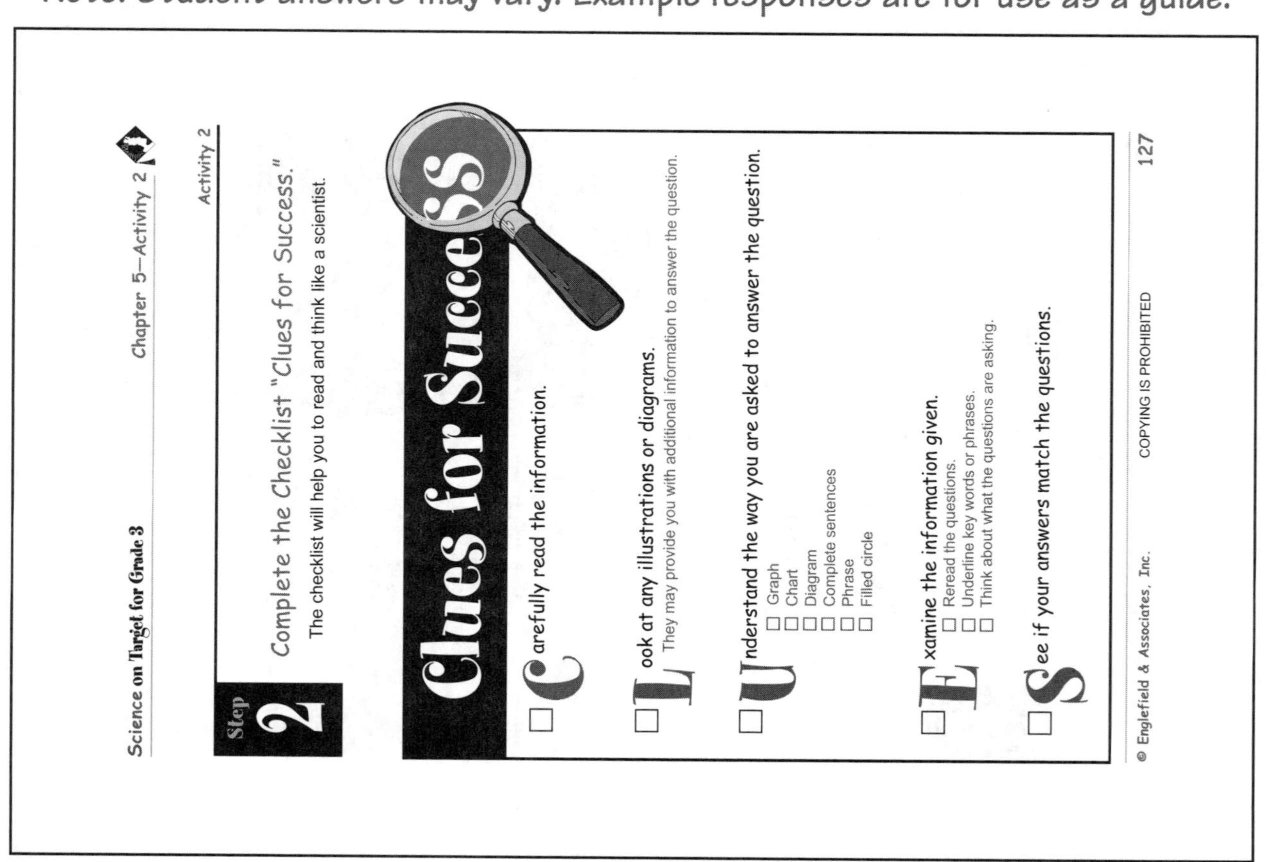
Science on Target for Grade 3 Chapter 5—Activity 2

Activity 2

Step 2

Complete the Checklist "Clues for Success."

The checklist will help you to read and think like a scientist.

Clues for Success

☐ **C**arefully read the information.

☐ **L**ook at any illustrations or diagrams.
They may provide you with additional information to answer the question.

☐ **U**nderstand the way you are asked to answer the question.
☐ Graph
☐ Chart
☐ Diagram
☐ Complete sentences
☐ Phrase
☐ Filled circle

☐ **E**xamine the information given.
☐ Reread the questions.
☐ Underline key words or phrases.
☐ Think about what the questions are asking.

☐ **S**ee if your answers match the questions.

© Englefield & Associates, Inc. COPYING IS PROHIBITED 127

Activity 2

Step 4

Answer the following questions for "It's a Bonus: Science Sometimes Leads to Bonus Results!" using information from your graphic organizer.

NASA scientists used products created for use in everyday life in the outer space missions. Two everyday products that are often linked to space explorations are Tang, a powdered orange drink, and the cordless drill (a drill that is battery operated).

1. Why do you think Tang, a powdered orange drink, was used for space flight?

   Tang was used because it was lighter/easier to take into space than liquid juice.

2. Why do you think it would be important to have a cordless drill to make repairs outside the space craft?

   Cordless drills were important because there is no electricity outside of the spacecraft.

Note: Student answers may vary. Example responses are for use as a guide.

3. Does the scenario show how information in different areas of science are linked together?

   ● yes ○ no

4. Using the information about NASA and bonus results, give **two** examples to support your answer.

   1. *Products were developed to use in space missions like foam for seat custions.*

   2. *Products were changed into things that people could use in their daily lives, like foam for mattresses.*

Chapter 5—Activity 3 Science on Target for Grade 3

Activity 3

Step 2

Complete the Checklist "Clues for Success."

The checklist will help you to read and think like a scientist.

Clues for Success

- ☐ **C**arefully read the information.
- ☐ **L**ook at any illustrations or diagrams.
  They may provide you with additional information to answer the question.
- ☐ **U**nderstand the way you are asked to answer the question.
  - ☐ Graph
  - ☐ Chart
  - ☐ Diagram
  - ☐ Complete sentences
  - ☐ Phrase
  - ☐ Filled circle
- ☐ **E**xamine the information given.
  - ☐ Reread the questions.
  - ☐ Underline key words or phrases.
  - ☐ Think about what the questions are asking.
- ☐ **S**ee if your answers match the questions.

132 COPYING IS PROHIBITED © Englefield & Associates, Inc.

Note: Student answers may vary. Example responses are for use as a guide.

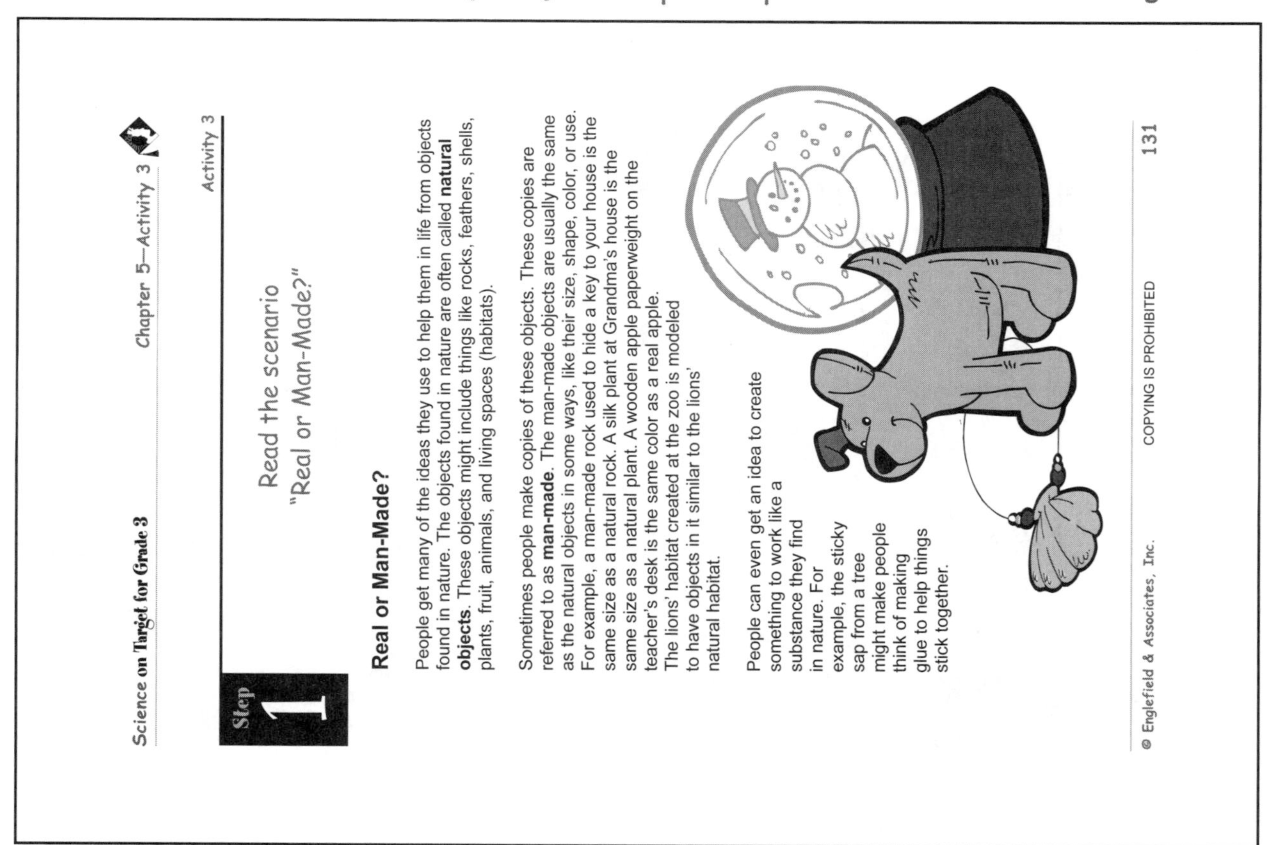
Science on Target for Grade 3 Chapter 5—Activity 3

Activity 3

Step 1

Read the scenario "Real or Man-Made?"

Real or Man-Made?

People get many of the ideas they use to help them in life from objects found in nature. The objects found in nature are often called **natural objects**. These objects might include things like rocks, feathers, shells, plants, fruit, animals, and living spaces (habitats).

Sometimes people make copies of these objects. These copies are referred to as **man-made**. The man-made objects are usually the same as the natural objects in some ways, like their size, shape, color, or use. For example, a man-made rock used to hide a key to your house is the same size as a natural rock. A silk plant at Grandma's house is the same size as a natural plant. A wooden apple paperweight on the teacher's desk is the same color as a real apple. The lions' habitat created at the zoo is modeled to have objects in it similar to the lions' natural habitat.

People can even get an idea to create something to work like a substance they find in nature. For example, the sticky sap from a tree might make people think of making glue to help things stick together.

© Englefield & Associates, Inc. COPYING IS PROHIBITED 131

Step 4

Answer the following questions for "Real or Man-Made?" using information from your graphic organizer.

A naturalist (a person who studies nature) brought a model of a raccoon's skeleton to class. It was about 10 inches high and 15 inches long, white in color, and the bones were held together with wires.

Consider the **four** characteristics of a man-made object from the scenario. Use a completely filled circle to show if the skeleton models that characteristic. Then, use a complete sentence telling why you think the skeleton models that characteristic.

1. Is the man-made skeleton the same **size** as a natural skeleton?

● yes ○ no

Why?

A raccoon is about 10 inches high and 15 inches long. This is about the size of a real raccoon.

Note: Student answers may vary. Example responses are for use as a guide.

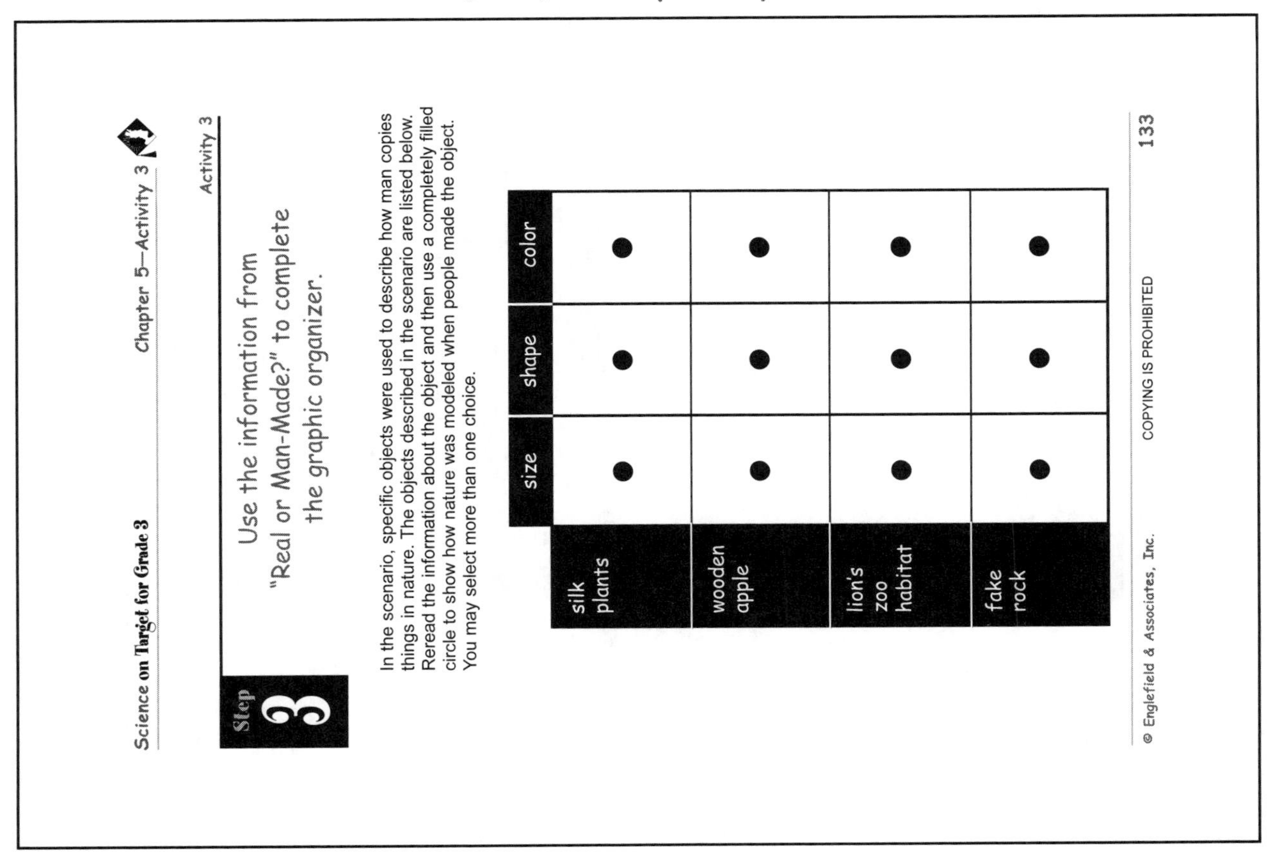

Step 3

Use the information from "Real or Man-Made?" to complete the graphic organizer.

In the scenario, specific objects were used to describe how man copies things in nature. The objects described in the scenario are listed below. Reread the information about the object and then use a completely filled circle to show how nature was modeled when people made the object. You may select more than one choice.

	size	shape	color
silk plants	●	●	●
wooden apple	●	●	●
lion's zoo habitat	●	●	●
fake rock	●	●	●

Note: Student answers may vary. Example responses are for use as a guide.

2. Does the man-made skeleton have the same **shape** as a natural skeleton?

● yes   ○ no

Why?

The bones are the same size and in the same place as in a real raccoon skeleton.

3. Is the man-made skeleton the same **color** as a natural skeleton?

○ yes   ● no

Why?

Bones are not white in real animals.

4. Is the naturalist's use of the man-made skeleton the way a skeleton is used in nature?

○ yes   ● no

Why?

The naturalist is showing the class what a raccoon skeleton looks like. In real life, the bones give the raccoon its shape.

Chapter 6

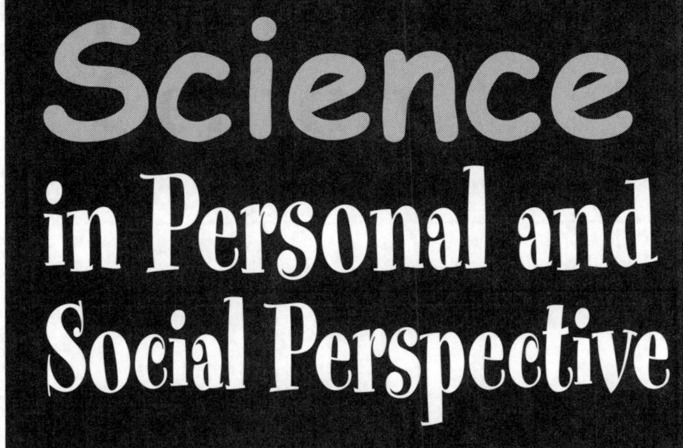

Chapter 6 of the *Science on Target for Grade 3 Student Workbook*, covers the National Content Standards for Science in Personal and Social Perspective. The standards are as follows:

Science in Personal and Social Perspective

- Personal health.
- Characteristics and changes in populations.
- Types of resources.
- Changes in environments.
- Science and technology in local challenges.

All of the pages from Chapter 6 of the *Science on Target for Grade 3 Student Workbook*, are reproduced in this Parent/Teacher Edition in reduced-page format with sample answers. These activities will help your students develop the skills necessary to understand Science in Personal and Social Perspective.

Students should use the "Clues for Success" Checklist, for each activity in this section, as a tool to help them do their best work.

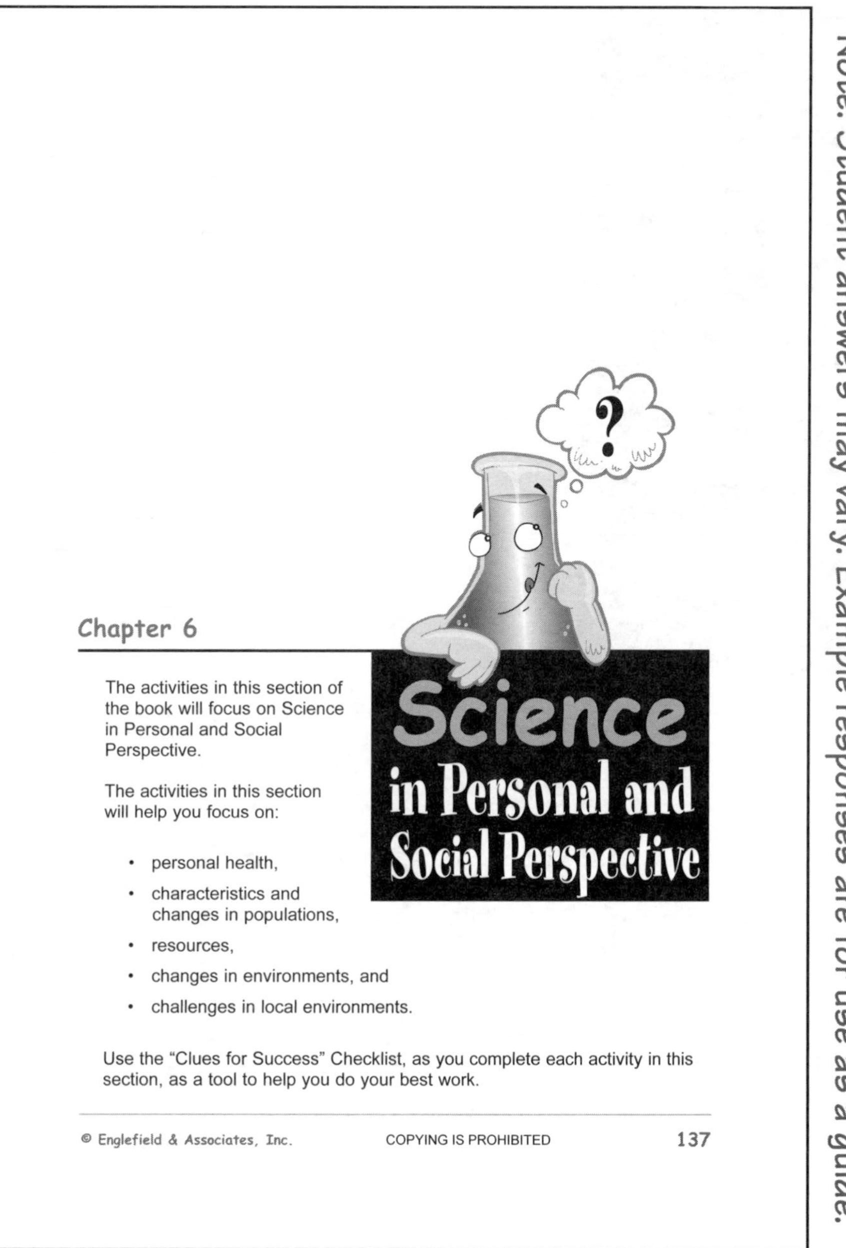

Chapter 6

Science in Personal and Social Perspective

The activities in this section of the book will focus on Science in Personal and Social Perspective.

The activities in this section will help you focus on:

- personal health,
- characteristics and changes in populations,
- resources,
- changes in environments, and
- challenges in local environments.

Use the "Clues for Success" Checklist, as you complete each activity in this section, as a tool to help you do your best work.

Note: Student answers may vary. Example responses are for use as a guide.

Activity 1

Step 1 Read the scenario "Lightning Flash!"

Lightning Flash!

Miranda was planning to watch her sister run in the district track meet. She was worried for her safety because the weather channel said that conditions for lightning were in the area. Miranda's teacher agreed that lightning during a thunderstorm caused people injuries. The teacher assured Miranda that she would get more information, but the athletic director would make decisions to keep everyone safe.

The athletic director said he would check the weather report for the day from a NOAA (National Oceanic and Atmospheric Administration) weather radio. This is a tool for getting information about the daily forecasts and approaching storms.

In addition, the athletic director would also use a hand-held lightning detector. The lightning detector is a battery operated, hand-held instrument with an electronic system that can pick up lightning (electrical) activity as far as 40 miles away. It also tracks the storm as it gets closer. The athletic director said that it gives off a warning tone that he can hear and lights up to show him how far away the last and closest lightning was seen. This will give the teams and spectators time to leave the track if lightning is approaching.

The teacher told Miranda that people always worried about lightning strikes. In the mid 1770's, Benjamin Franklin wanted to protect people and buildings from lightning. Franklin invented lightning rods to attach to building rooftops. When lightning strikes the rod, its energy is conducted to the ground instead of passing through the building, where it could start a fire. These rods provide the lightning with a quick path to the ground. Lightning rods are still used on buildings today.

Miranda's teacher ended their conversation with a little jingle for Miranda to remember: *"If you hear it (thunder), clear it! If you see it (lightning), flee it!"*

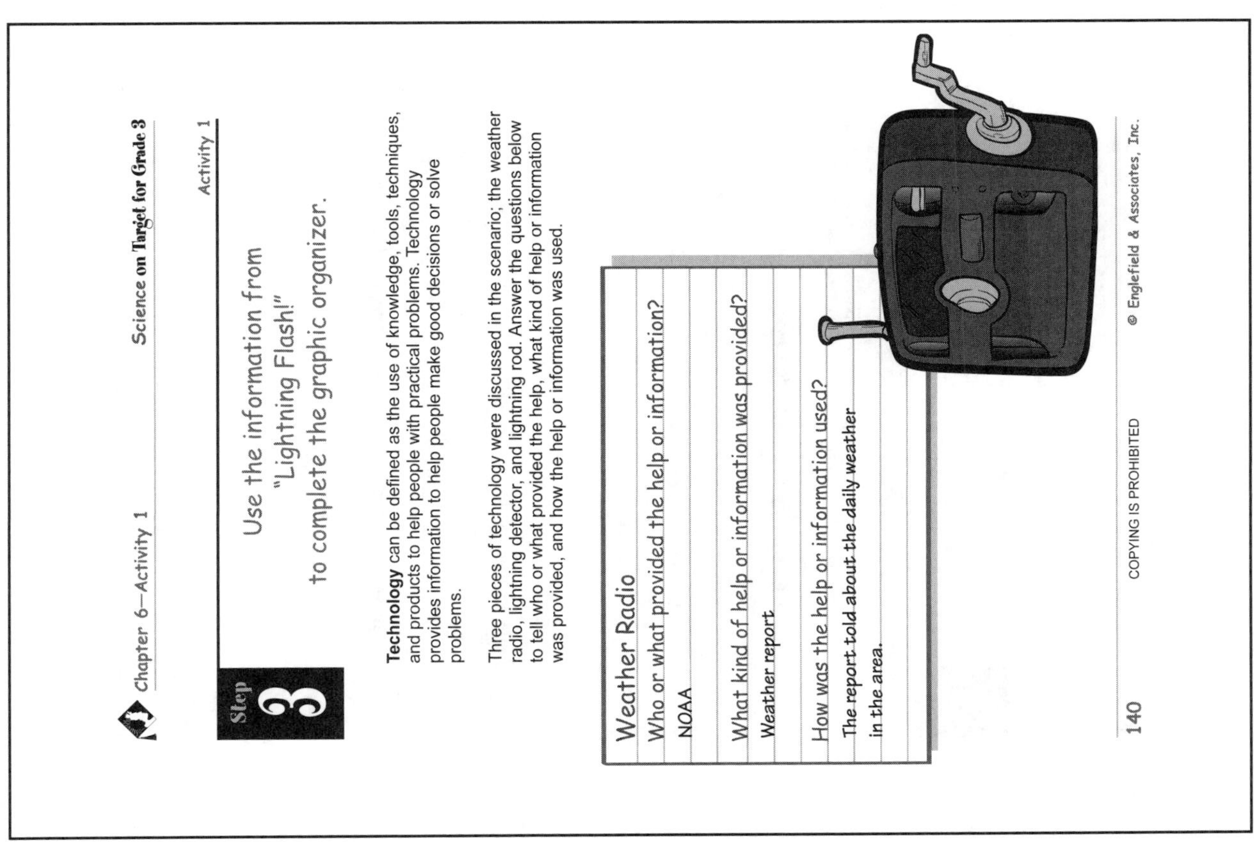

Chapter 6—Activity 1 Science on Target for Grade 3

Activity 1

Step 3

Use the information from "Lightning Flash!" to complete the graphic organizer.

Technology can be defined as the use of knowledge, tools, techniques, and products to help people with practical problems. Technology provides information to help people make good decisions or solve problems.

Three pieces of technology were discussed in the scenario; the weather radio, lightning detector, and lightning rod. Answer the questions below to tell who or what provided the help, what kind of help or information was provided, and how the help or information was used.

Weather Radio

Who or what provided the help or information?

NOAA

What kind of help or information was provided?

Weather report

How was the help or information used?

The report told about the daily weather in the area.

140 COPYING IS PROHIBITED © Englefield & Associates, Inc.

Note: Student answers may vary. Example responses are for use as a guide.

Science on Target for Grade 3 Chapter 6—Activity 1

Activity 1

Step 2

Complete the Checklist "Clues for Success."

The checklist will help you to read and think like a scientist.

Clues for Success

☐ **C**arefully read the information.

☐ **L**ook at any illustrations or diagrams.
They may provide you with additional information to answer the question.

☐ **U**nderstand the way you are asked to answer the question.
- ☐ Graph
- ☐ Chart
- ☐ Diagram
- ☐ Complete sentences
- ☐ Phrase
- ☐ Filled circle

☐ **E**xamine the information given.
- ☐ Reread the questions.
- ☐ Underline key words or phrases.
- ☐ Think about what the questions are asking.

☐ **S**ee if your answers match the questions.

© Englefield & Associates, Inc. COPYING IS PROHIBITED 139

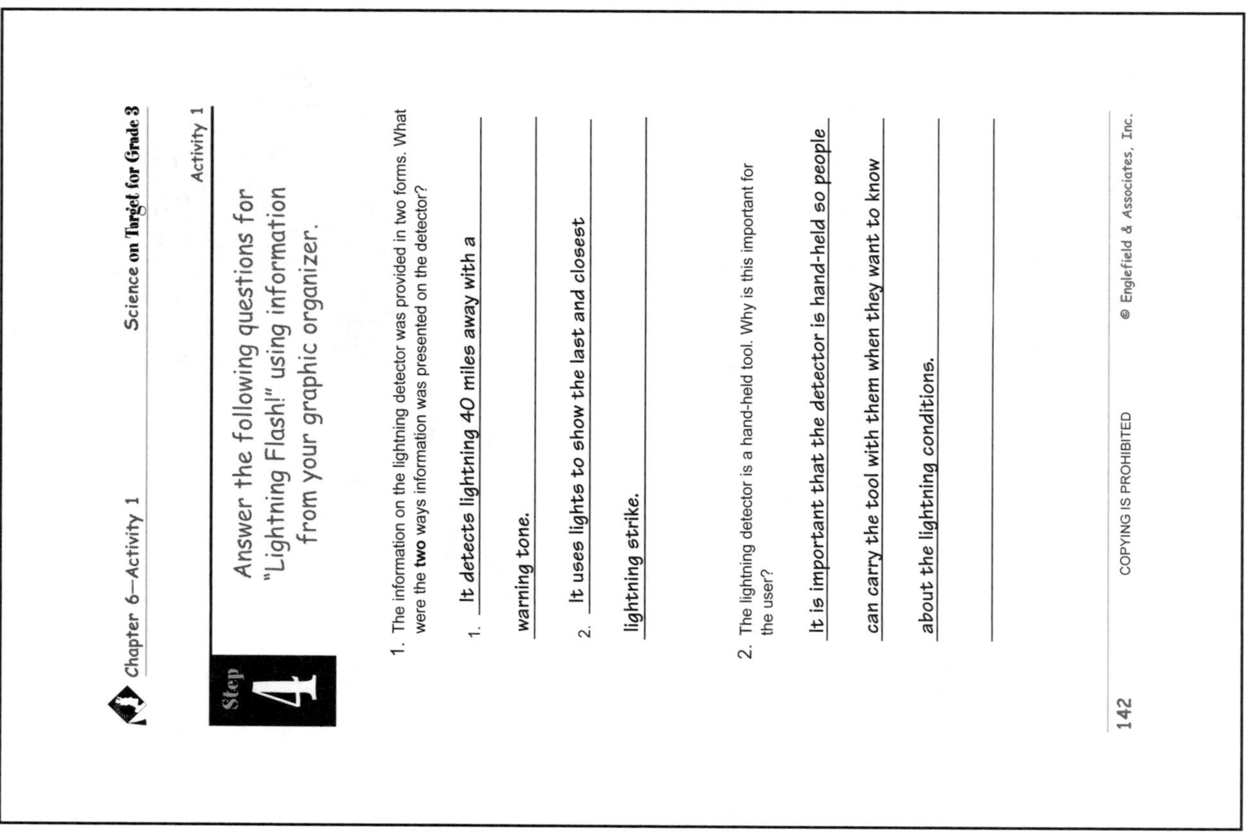
Chapter 6—Activity 1 Science on Target for Grade 3

Step 4 Answer the following questions for "Lightning Flash!" using information from your graphic organizer.

Activity 1

1. The information on the lightning detector was provided in two forms. What were the **two** ways information was presented on the detector?

1. It detects lightning 40 miles away with a warning tone.

2. It uses lights to show the last and closest lightning strike.

2. The lightning detector is a hand-held tool. Why is this important for the user?

It is important that the detector is hand-held so people can carry the tool with them when they want to know about the lightning conditions.

142 COPYING IS PROHIBITED © Englefield & Associates, Inc.

Note: Student answers may vary. Example responses are for use as a guide.

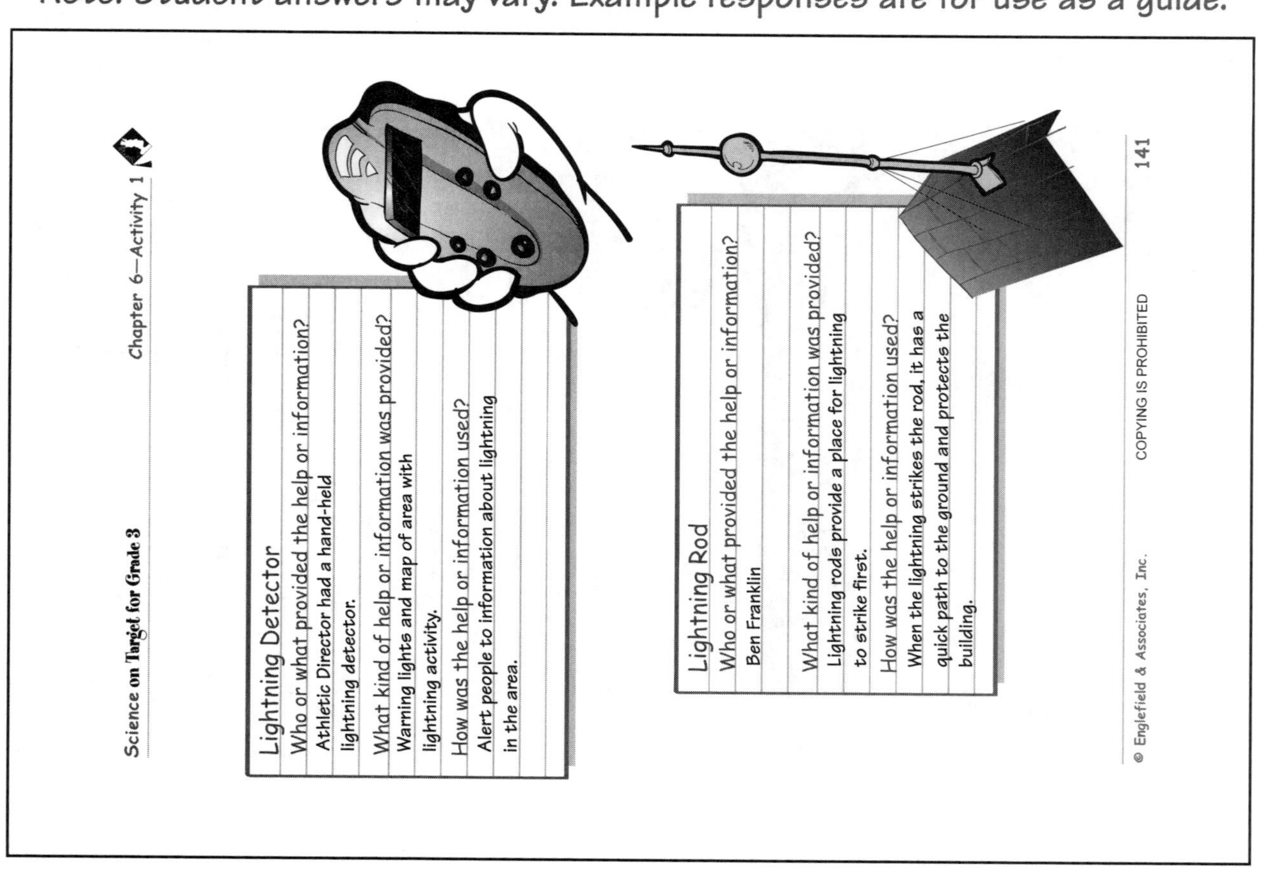
Science on Target for Grade 3 Chapter 6—Activity 1

Lightning Detector

Who or what provided the help or information?

Athletic Director had a hand-held lightning detector.

What kind of help or information was provided?

Warning lights and map of area with lightning activity.

How was the help or information used?

Alert people to information about lightning in the area.

Lightning Rod

Who or what provided the help or information?

Ben Franklin

What kind of help or information was provided?

Lightning rods provide a place for lightning to strike first.

How was the help or information used?

When the lightning strikes the rod, it has a quick path to the ground and protects the building.

© Englefield & Associates, Inc. COPYING IS PROHIBITED 141

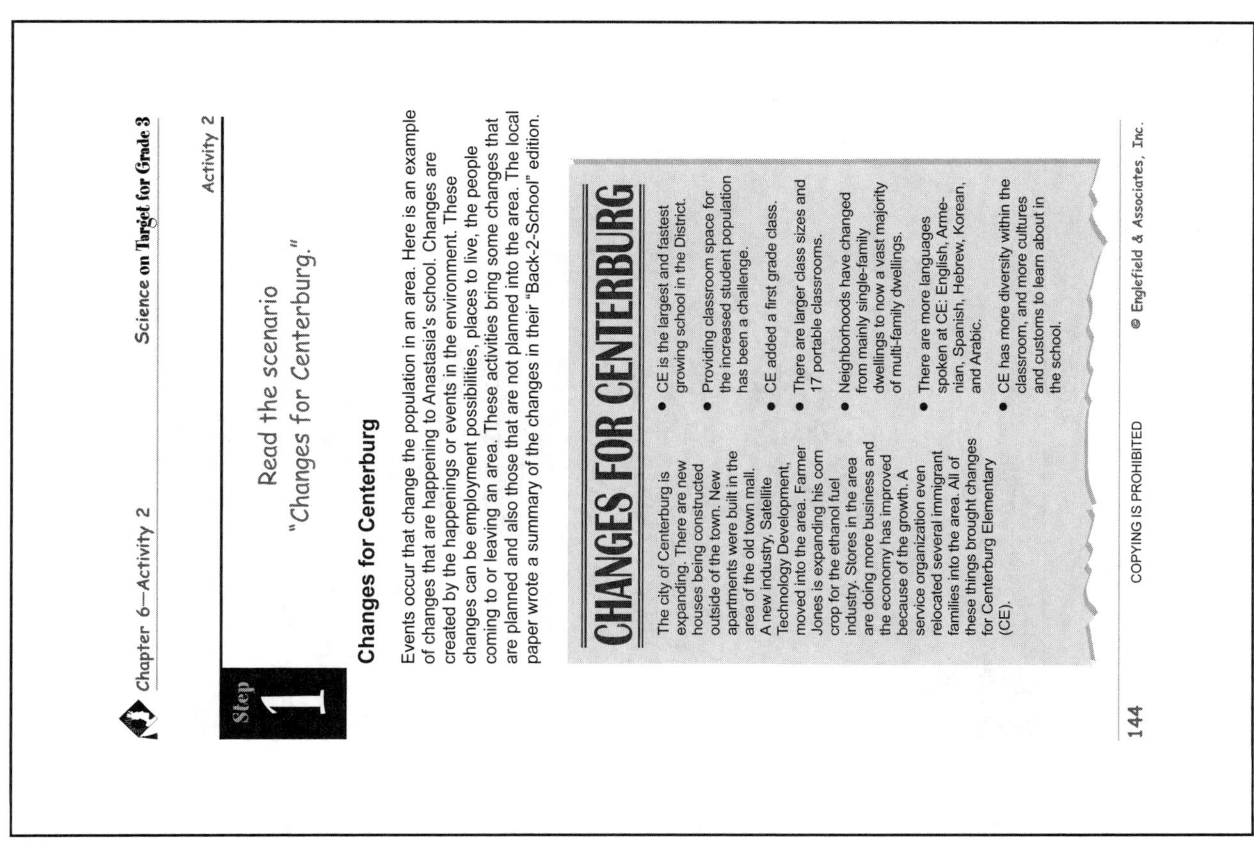

Activity 2

Step 1

Read the scenario
"Changes for Centerburg."

Changes for Centerburg

Events occur that change the population in an area. Here is an example of changes that are happening to Anastasia's school. Changes are created by the happenings or events in the environment. These changes can be employment possibilities, places to live, the people coming to or leaving an area. These activities bring some changes that are planned and also those that are not planned into the area. The local paper wrote a summary of the changes in their "Back-2-School" edition.

CHANGES FOR CENTERBURG

The city of Centerburg is expanding. There are new houses being constructed outside of the town. New apartments were built in the area of the old town mall. A new industry, Satellite Technology Development, moved into the area. Farmer Jones is expanding his corn crop for the ethanol fuel industry. Stores in the area are doing more business and the economy has improved because of the growth. A service organization even relocated several immigrant families into the area. All of these things brought changes for Centerburg Elementary (CE).

- CE is the largest and fastest growing school in the District.
- Providing classroom space for the increased student population has been a challenge.
- CE added a first grade class.
- There are larger class sizes and 17 portable classrooms.
- Neighborhoods have changed from mainly single-family dwellings to now a vast majority of multi-family dwellings.
- There are more languages spoken at CE: English, Armenian, Spanish, Hebrew, Korean, and Arabic.
- CE has more diversity within the classroom, and more cultures and customs to learn about in the school.

Note: Student answers may vary. Example responses are for use as a guide.

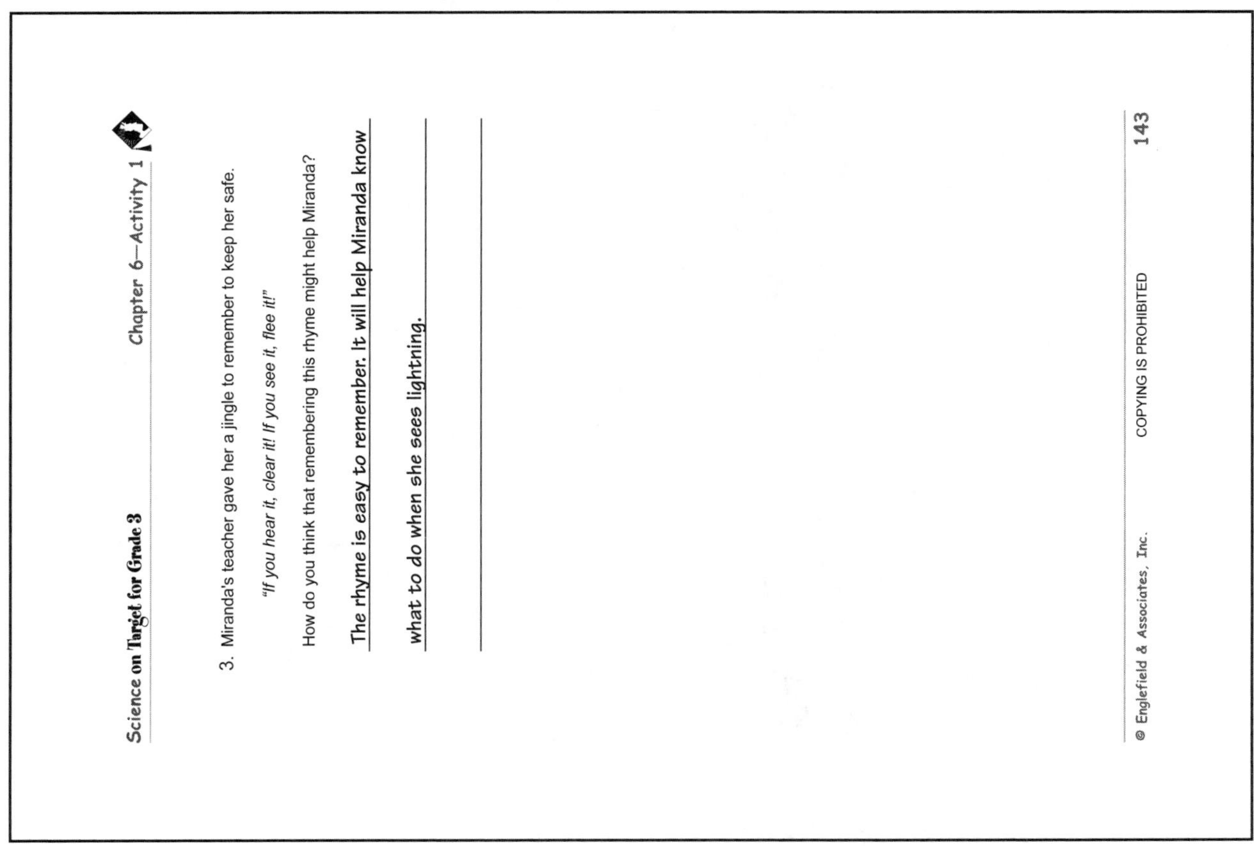

3. Miranda's teacher gave her a jingle to remember to keep her safe.

   "If you hear it, clear it! If you see it, flee it!"

   How do you think that remembering this rhyme might help Miranda?

   The rhyme is easy to remember. It will help Miranda know what to do when she sees lightning.

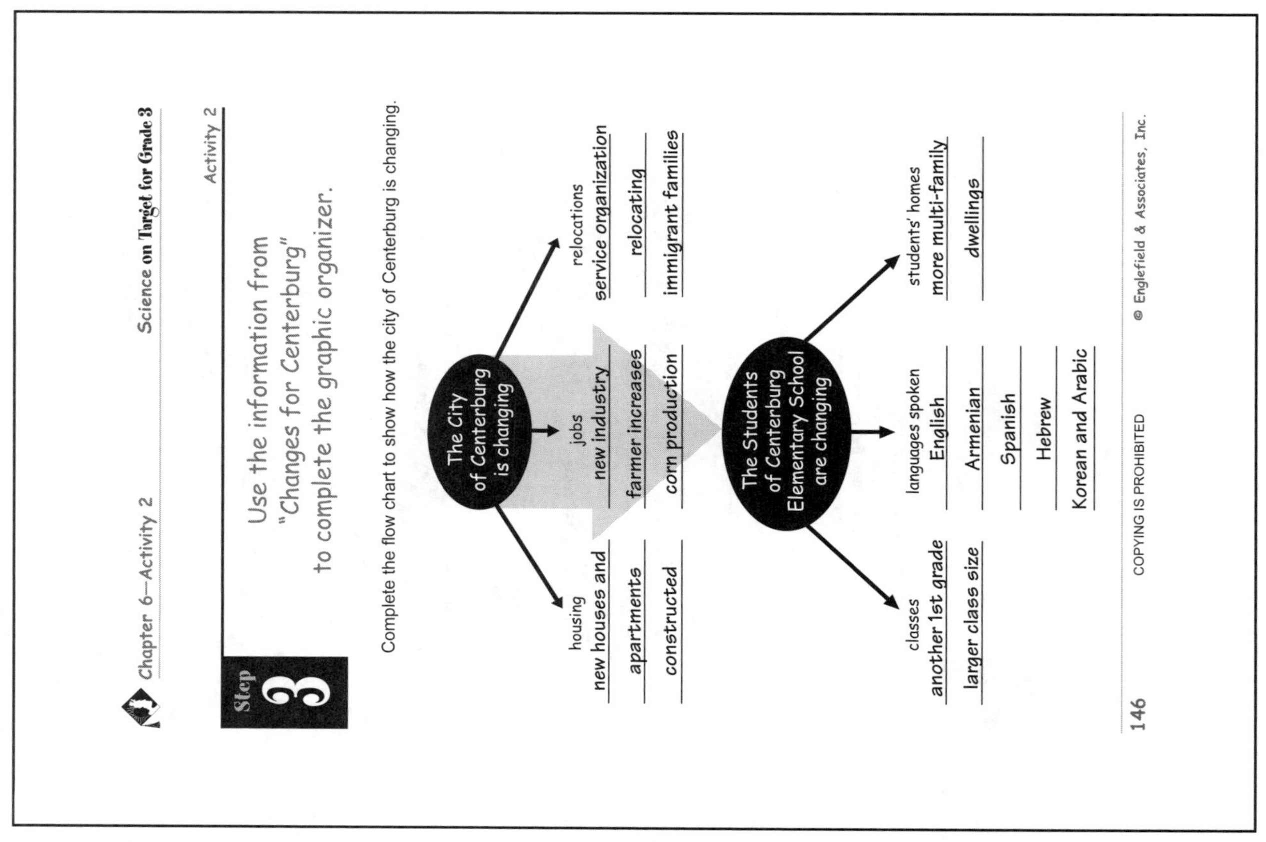
Chapter 6—Activity 2 Science on Target for Grade 3

Activity 2

Step 3 Use the information from "Changes for Centerburg" to complete the graphic organizer.

Complete the flow chart to show how the city of Centerburg is changing.

146 COPYING IS PROHIBITED © Englefield & Associates, Inc.

Note: Student answers may vary. Example responses are for use as a guide.

Science on Target for Grade 3 Chapter 6—Activity 2

Activity 2

Step 2 Complete the Checklist "Clues for Success."

The checklist will help you to read and think like a scientist.

Clues for Success

- ☐ **C**arefully read the information.
- ☐ **L**ook at any illustrations or diagrams.
  They may provide you with additional information to answer the question.
- ☐ **U**nderstand the way you are asked to answer the question.
  - ☐ Graph
  - ☐ Chart
  - ☐ Diagram
  - ☐ Complete sentences
  - ☐ Phrase
  - ☐ Filled circle
- ☐ **E**xamine the information given.
  - ☐ Reread the questions.
  - ☐ Underline key words or phrases.
  - ☐ Think about what the questions are asking.
- ☐ **S**ee if your answers match the questions.

© Englefield & Associates, Inc. COPYING IS PROHIBITED 145

Activity 2

Step 4

Answer the following questions for "Changes for Centerburg" using information from your graphic organizer.

The changes to the population in the city of Centerburg is also changing the natural environment around the city. Some of these changes are helpful and some are harmful to the people and the surrounding natural areas. Be sure to use a complete sentence to explain your answer choice.

1. More of the natural areas are being used for parking.

   ○ helpful change ● harmful change

   Why?

   Animals and plants will lose their natural habitat.

2. The trash is increasing.

   ○ helpful change ● harmful change

   Why?

   More litter will be created. It is an eyesore and will attract mice and rats.

3. More people are using the library.

   ● helpful change ○ harmful change

   Why?

   The library will serve more people. It may create more jobs, buy more books, and provide more services like story time.

4. The class sizes are getting bigger at Centerburg Elementary.

   ○ helpful change ● harmful change

   Why?

   The class size may change the attention each student receives from the teacher.

Note: Student answers may vary. Example responses are for use as a guide.

Think about this change:

People are using the parks for more cultural celebrations.

5. Write a complete sentence to describe why this is a helpful change to the community.

The people in the area will learn about different customs.

6. Write a complete sentence to describe why this is a harmful change to the community.

The celebration will produce more trash in the parks.

People not involved in the celebrations may not be able to enjoy the park.

Note: Student answers may vary. Example responses are for use as a guide.

Activity 3

Step 1 Read the scenario "Canada Geese Change in Migration Patterns."

Canada Geese Change in Migration Patterns

Canada Geese were taking over the field near school. They traveled onto the school yard making loud noises and leaving droppings everywhere. Sometimes they even stopped traffic in the travel route to the school door.

The park naturalist came to the school to help the students learn more about the increase in the number of Canada Geese in their area. The class was surprised to learn that the population of Canada Geese was very low from the 1800s to the 1900s because they were hunted too much. They also lost the environment they needed for life. By 1952, scientists thought the Canada Geese were extinct (no longer living).

In 1962, a naturalist found a few Canada Geese when he was conducting a wildlife survey. Many changes took place to protect the Canada Geese including laws to protect them from hunters and their natural predators including foxes, coyotes, wolves, and bears. Naturalist also developed programs to create the habitat Canada Geese need to live.

With that background information, he showed a presentation with these points:

- A field provides the ideal habitat for Canada Geese—a flat, green, grassy area with a pond and food (plants).
- Canada Geese are not bothered by people.
- The area has very few of their natural predators.
- Some people bring them food.
- Canada Geese remember and return to areas that provide their needs for living.

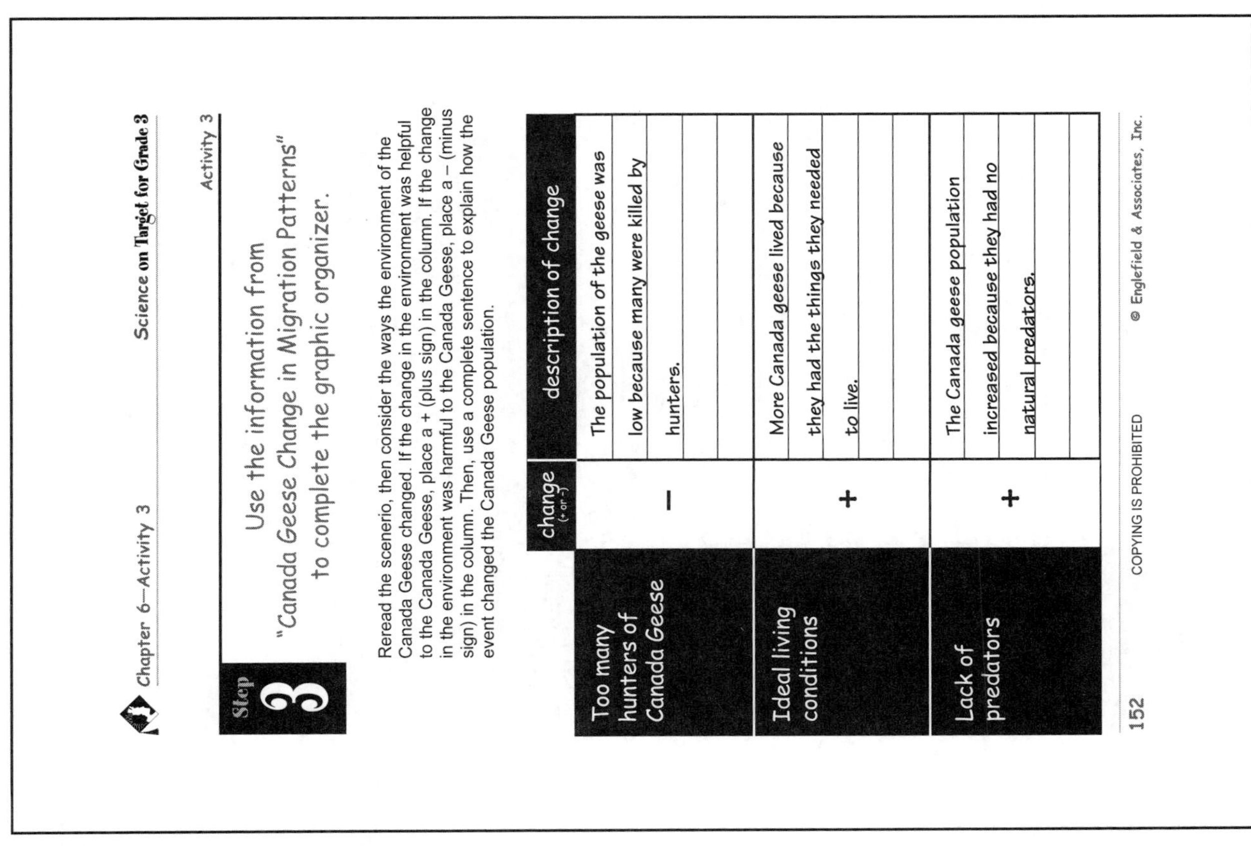

Activity 3

Step 3 Use the information from "*Canada Geese Change in Migration Patterns*" to complete the graphic organizer.

Reread the scenerio, then consider the ways the environment of the Canada Geese changed. If the change in the environment was helpful to the Canada Geese, place a + (plus sign) in the column. If the change in the environment was harmful to the Canada Geese, place a – (minus sign) in the column. Then, use a complete sentence to explain how the event changed the Canada Geese population.

	change (+ or –)	description of change
Too many hunters of Canada Geese	–	The population of the geese was low because many were killed by hunters.
Ideal living conditions	+	More Canada geese lived because they had the things they needed to live.
Lack of predators	+	The Canada geese population increased because they had no natural predators.

Note: Student answers may vary. Example responses are for use as a guide.

Activity 3

Step 2 Complete the Checklist "Clues for Success."

The checklist will help you to read and think like a scientist.

Clues for Success

- ☐ **C**arefully read the information.
- ☐ **L**ook at any illustrations or diagrams.
  They may provide you with additional information to answer the question.
- ☐ **U**nderstand the way you are asked to answer the question.
  - ☐ Graph
  - ☐ Chart
  - ☐ Diagram
  - ☐ Complete sentences
  - ☐ Phrase
  - ☐ Filled circle
- ☐ **E**xamine the information given.
  - ☐ Reread the questions.
  - ☐ Underline key words or phrases.
  - ☐ Think about what the questions are asking.
- ☐ **S**ee if your answers match the questions.

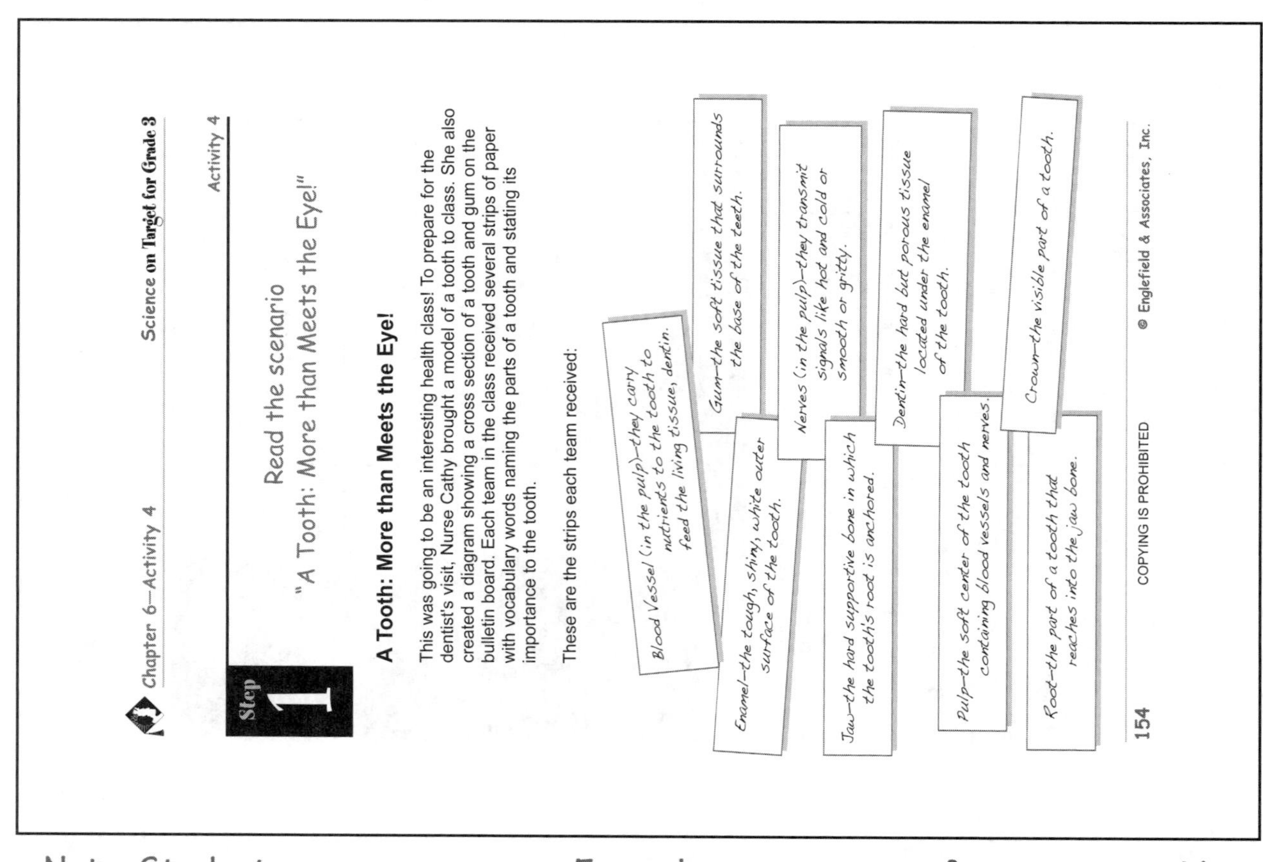
Chapter 6—Activity 4 Science on Target for Grade 3

Activity 4

Step 1

Read the scenario
"A Tooth: More than Meets the Eye!"

A Tooth: More than Meets the Eye!

This was going to be an interesting health class! To prepare for the dentist's visit, Nurse Cathy brought a model of a tooth to class. She also created a diagram showing a cross section of a tooth and gum on the bulletin board. Each team in the class received several strips of paper with vocabulary words naming the parts of a tooth and stating its importance to the tooth.

These are the strips each team received:

Blood Vessel (in the pulp)—they carry nutrients to the tooth to feed the living tissue, dentin.

Gum—the soft tissue that surrounds the base of the teeth.

Enamel—the tough, shiny, white outer surface of the tooth.

Nerves (in the pulp)—they transmit signals like hot and cold or smooth or gritty.

Jaw—the hard supportive bone in which the tooth's root is anchored.

Dentin—the hard but porous tissue located under the enamel of the tooth.

Pulp—the soft center of the tooth containing blood vessels and nerves.

Crown—the visible part of a tooth.

Root—the part of a tooth that reaches into the jaw bone.

154 COPYING IS PROHIBITED © Englefield & Associates, Inc.

Note: Student answers may vary. Example responses are for use as a guide.

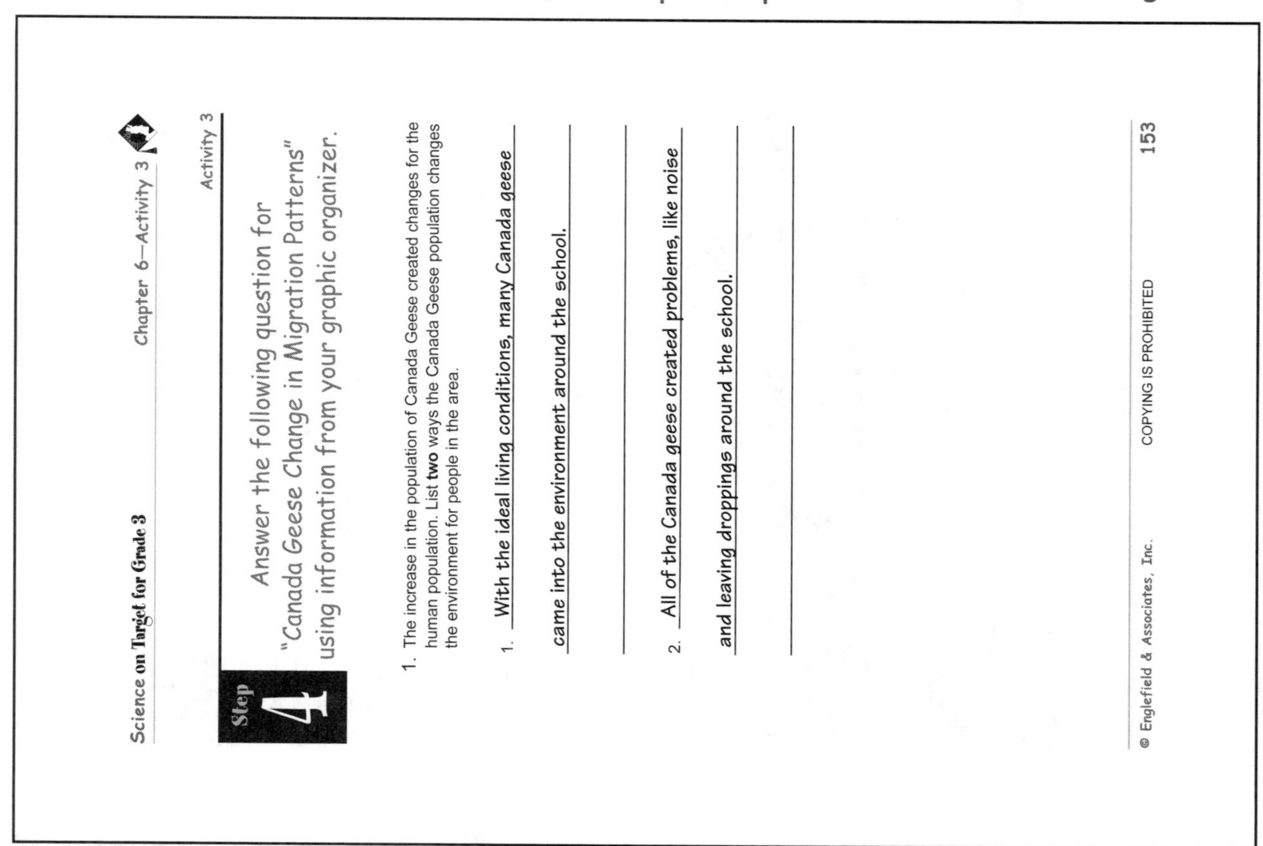
Science on Target for Grade 3 Chapter 6—Activity 3

Activity 3

Step 4

Answer the following question for
"Canada Geese Change in Migration Patterns"
using information from your graphic organizer.

1. The increase in the population of Canada Geese created changes for the human population. List **two** ways the Canada Geese population changes the environment for people in the area.

1. With the ideal living conditions, many Canada geese came into the environment around the school.

2. All of the Canada geese created problems, like noise and leaving droppings around the school.

© Englefield & Associates, Inc. COPYING IS PROHIBITED 153

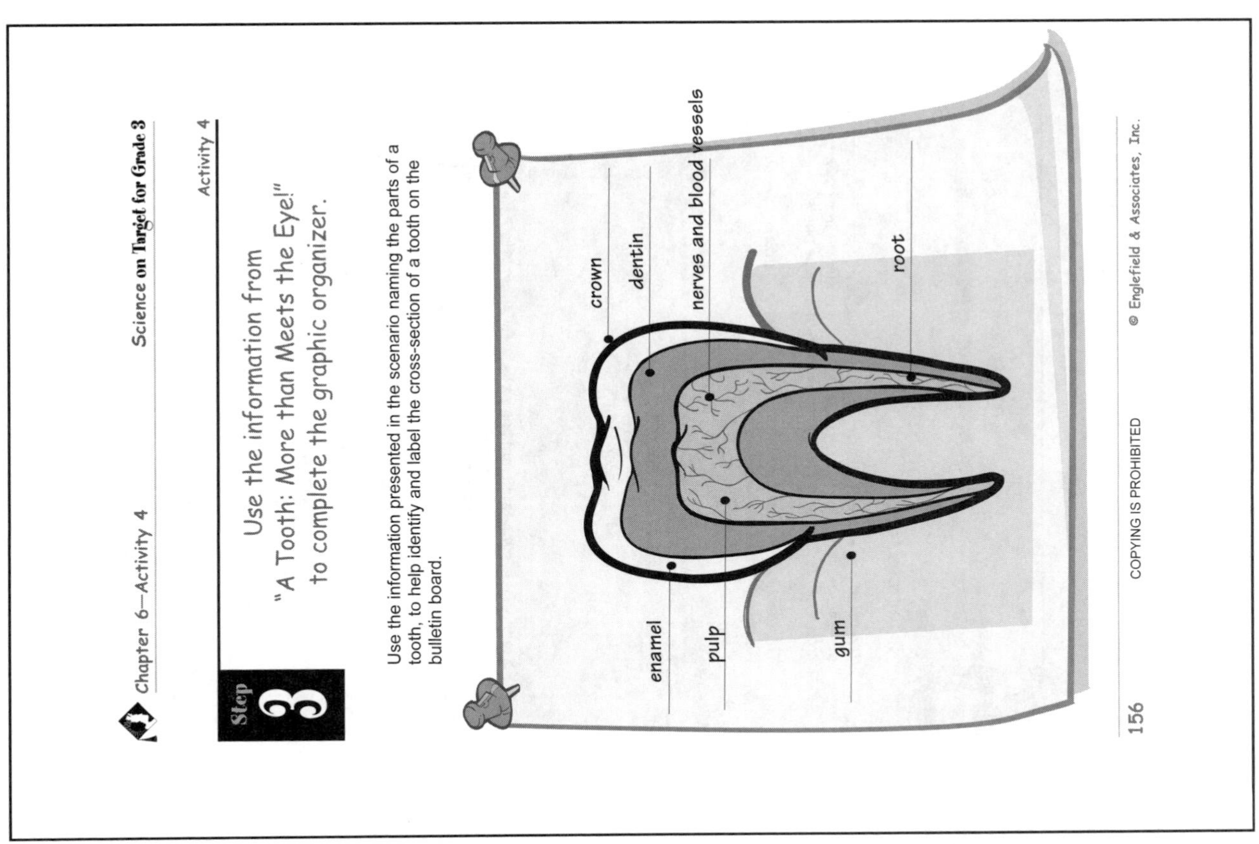

Chapter 6—Activity 4 Science on Target for Grade 3

Activity 4

Step 3

Use the information from "A Tooth: More than Meets the Eye!" to complete the graphic organizer.

Use the information presented in the scenario naming the parts of a tooth, to help identify and label the cross-section of a tooth on the bulletin board.

156 COPYING IS PROHIBITED © Englefield & Associates, Inc.

Note: Student answers may vary. Example responses are for use as a guide.

Science on Target for Grade 3 Chapter 6—Activity 4

Activity 4

Step 2

Complete the Checklist "Clues for Success."

The checklist will help you to read and think like a scientist.

Clues for Success

- ☐ **C**arefully read the information.
- ☐ **L**ook at any illustrations or diagrams.
  They may provide you with additional information to answer the question.
- ☐ **U**nderstand the way you are asked to answer the question.
  - ☐ Graph
  - ☐ Chart
  - ☐ Diagram
  - ☐ Complete sentences
  - ☐ Phrase
  - ☐ Filled circle
- ☐ **E**xamine the information given.
  - ☐ Reread the questions.
  - ☐ Underline key words or phrases.
  - ☐ Think about what the questions are asking.
- ☐ **S**ee if your answers match the questions.

© Englefield & Associates, Inc. COPYING IS PROHIBITED 155

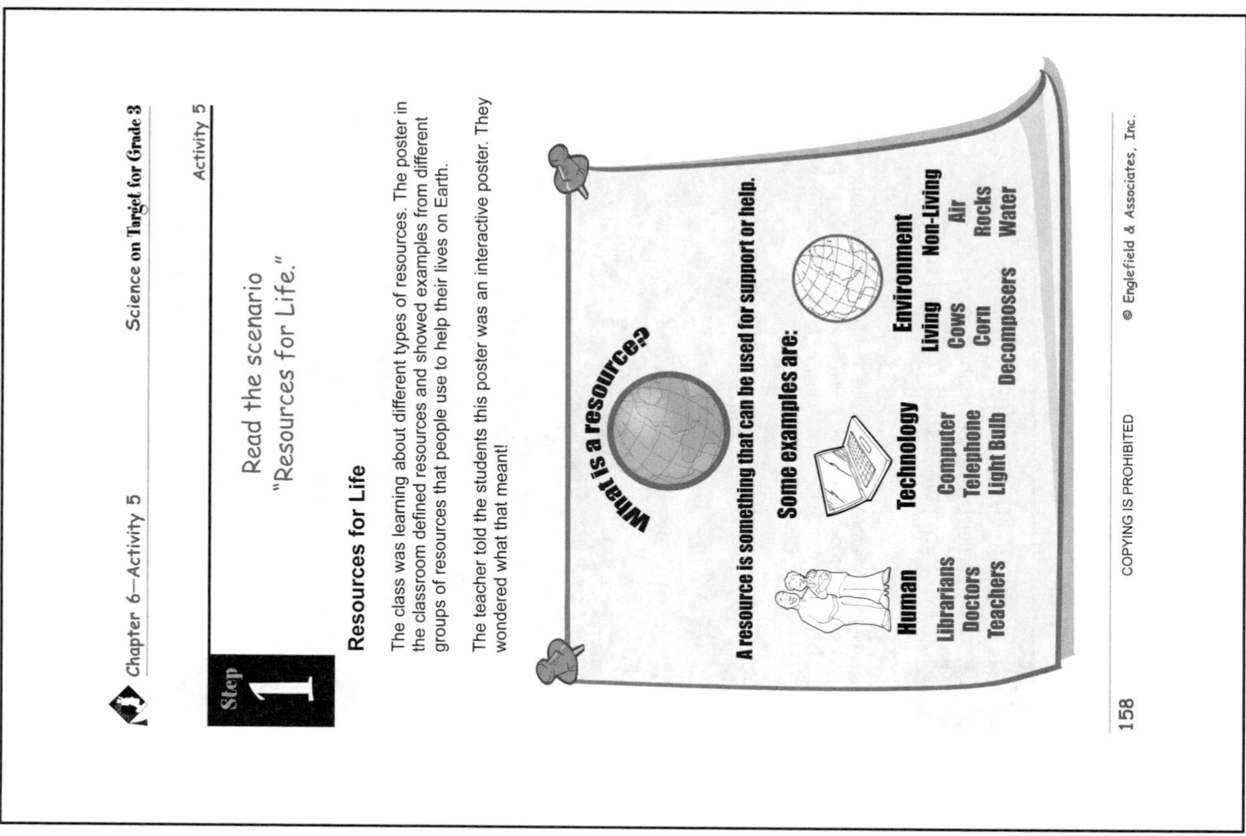

Chapter 6—Activity 5 Science on Target for Grade 3

Activity 5

Step 1

Read the scenario "Resources for Life."

Resources for Life

The class was learning about different types of resources. The poster in the classroom defined resources and showed examples from different groups of resources that people use to help their lives on Earth.

The teacher told the students this poster was an interactive poster. They wondered what that meant!

158 COPYING IS PROHIBITED © Englefield & Associates, Inc.

Note: Student answers may vary. Example responses are for use as a guide.

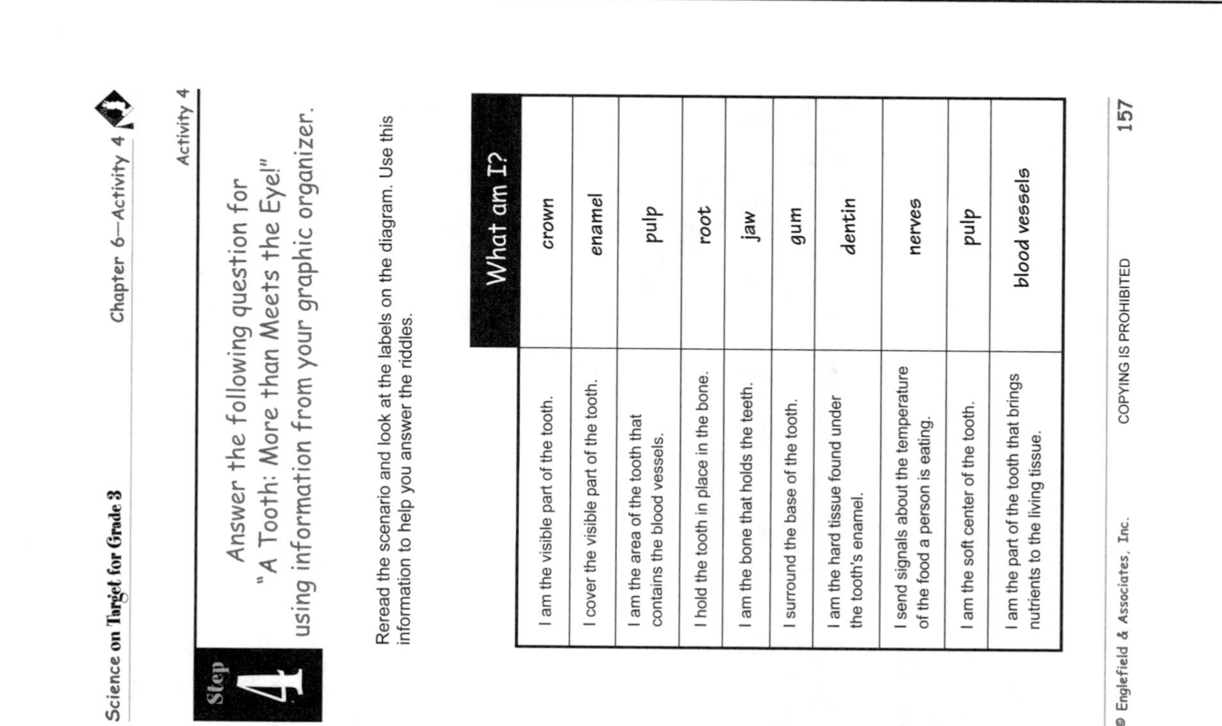

Science on Target for Grade 3 Chapter 6—Activity 4

Activity 4

Step 4

Answer the following question for "A Tooth: More than Meets the Eye!" using information from your graphic organizer.

Reread the scenario and look at the labels on the diagram. Use this information to help you answer the riddles.

	What am I?
I am the visible part of the tooth.	*crown*
I cover the visible part of the tooth.	*enamel*
I am the area of the tooth that contains the blood vessels.	*pulp*
I hold the tooth in place in the bone.	*root*
I am the bone that holds the teeth.	*jaw*
I surround the base of the tooth.	*gum*
I am the hard tissue found under the tooth's enamel.	*dentin*
I send signals about the temperature of the food a person is eating.	*nerves*
I am the soft center of the tooth.	*pulp*
I am the part of the tooth that brings nutrients to the living tissue.	*blood vessels*

© Englefield & Associates, Inc. COPYING IS PROHIBITED 157

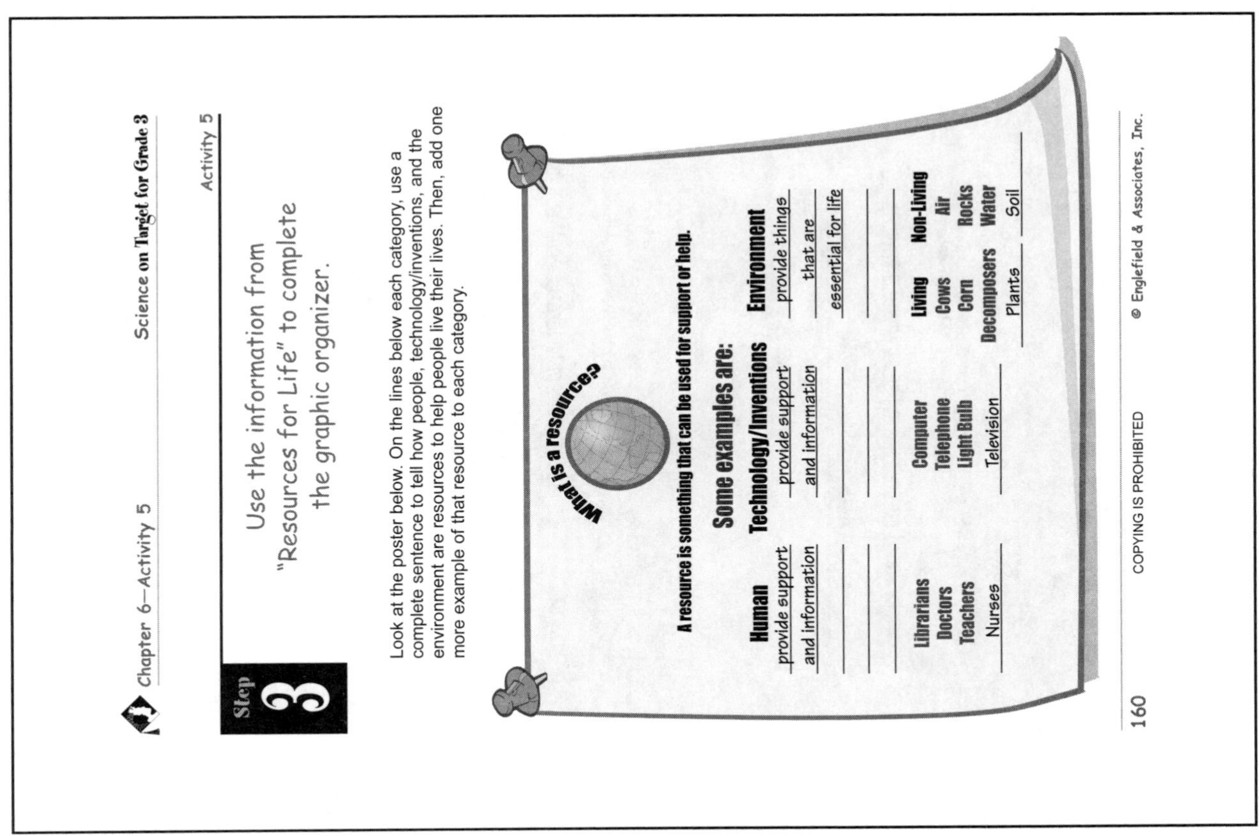

Chapter 6—Activity 5 Science on Target for Grade 3

Activity 5

Step 3

Use the information from "Resources for Life" to complete the graphic organizer.

Look at the poster below. On the lines below each category, use a complete sentence to tell how people, technology/inventions, and the environment are resources to help people live their lives. Then, add one more example of that resource to each category.

160 COPYING IS PROHIBITED © Englefield & Associates, Inc.

Note: Student answers may vary. Example responses are for use as a guide.

Science on Target for Grade 3 Chapter 6—Activity 5

Activity 5

Step 2

Complete the Checklist "Clues for Success."

The checklist will help you to read and think like a scientist.

Clues for Success

- ☐ **C**arefully read the information.
- ☐ **L**ook at any illustrations or diagrams.
  They may provide you with additional information to answer the question.
- ☐ **U**nderstand the way you are asked to answer the question.
  - ☐ Graph
  - ☐ Chart
  - ☐ Diagram
  - ☐ Complete sentences
  - ☐ Phrase
  - ☐ Filled circle
- ☐ **E**xamine the information given.
  - ☐ Reread the questions.
  - ☐ Underline key words or phrases.
  - ☐ Think about what the questions are asking.
- ☐ **S**ee if your answers match the questions.

© Englefield & Associates, Inc. COPYING IS PROHIBITED 159

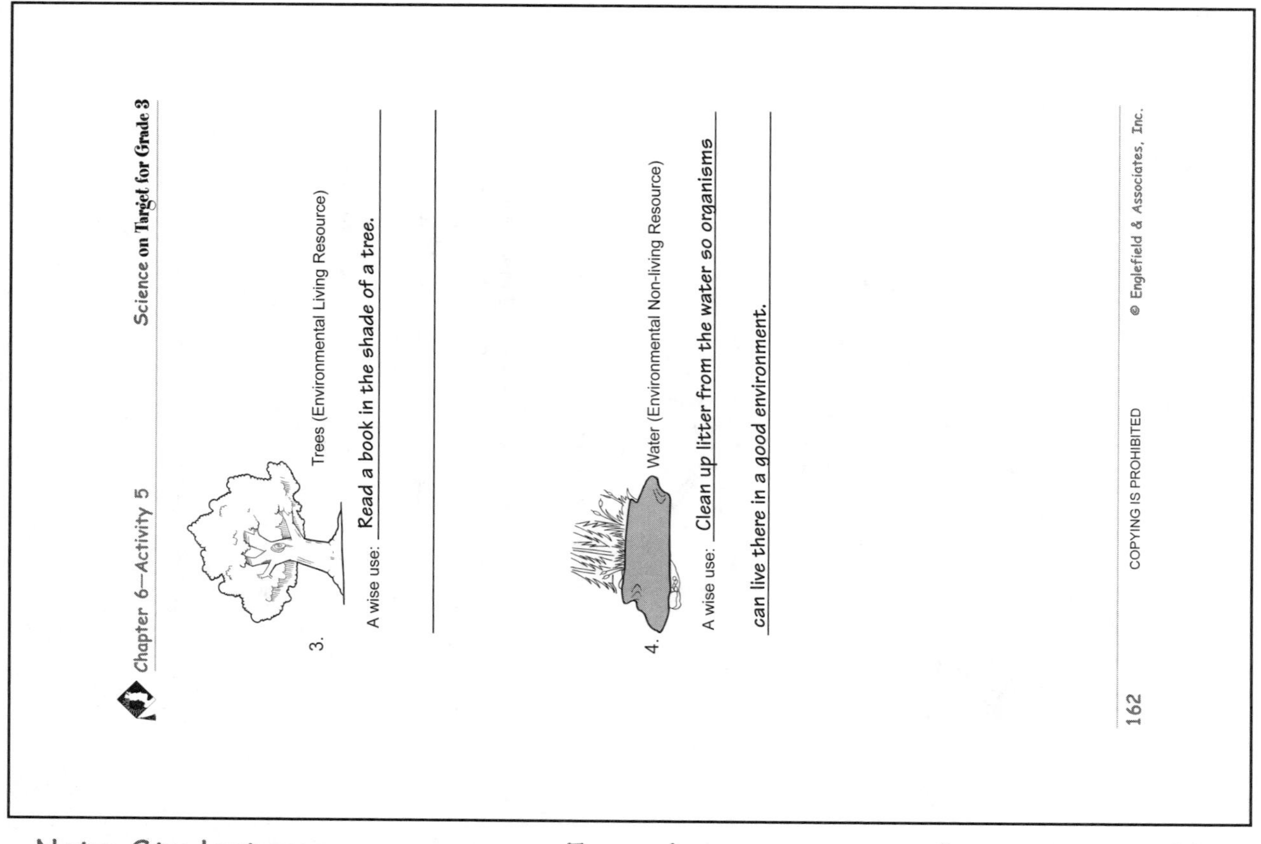

3. Trees (Environmental Living Resource)

A wise use: Read a book in the shade of a tree.

4. Water (Environmental Non-living Resource)

A wise use: Clean up litter from the water so organisms can live there in a good environment.

Note: Student answers may vary. Example responses are for use as a guide.

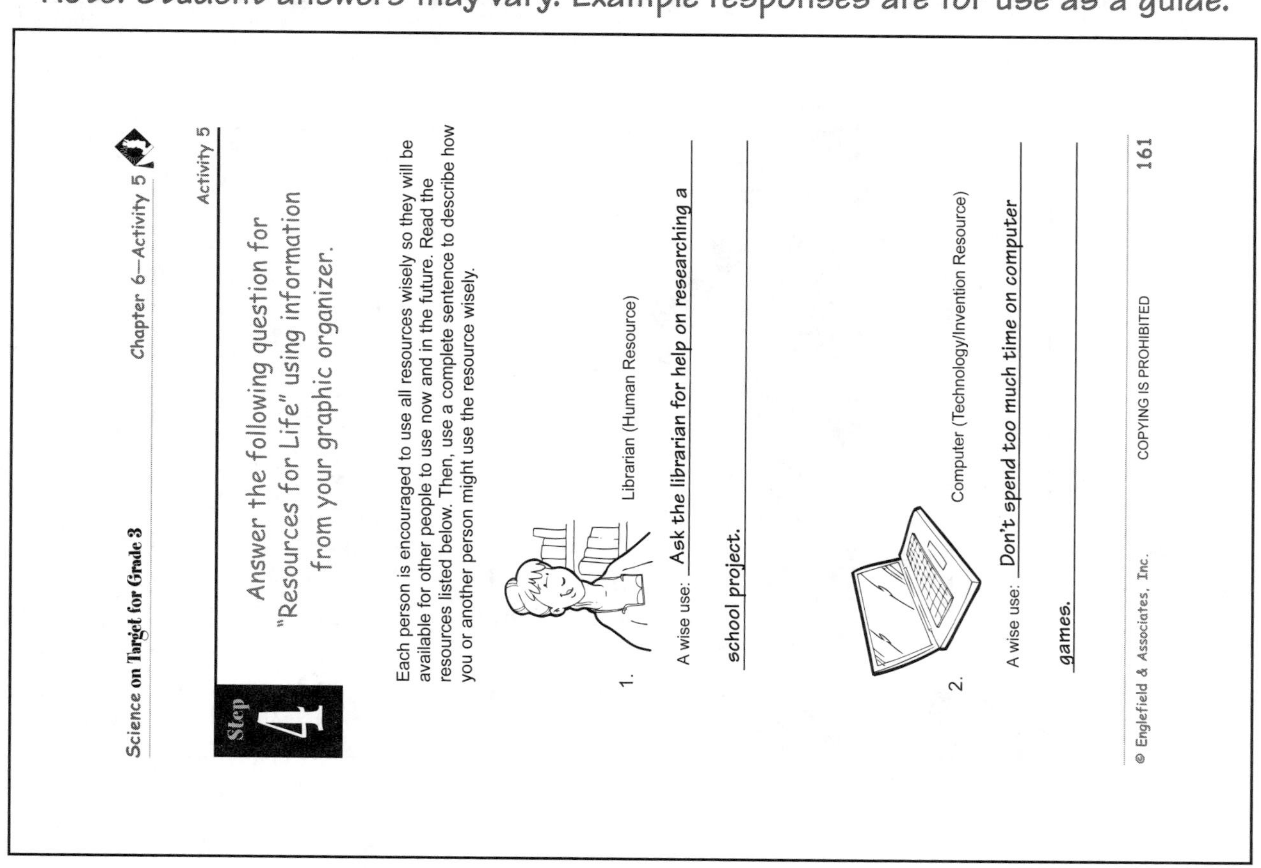

Activity 5

Step 4 Answer the following question for "Resources for Life" using information from your graphic organizer.

Each person is encouraged to use all resources wisely so they will be available for other people to use now and in the future. Read the resources listed below. Then, use a complete sentence to describe how you or another person might use the resource wisely.

1. Librarian (Human Resource)

A wise use: Ask the librarian for help on researching a school project.

2. Computer (Technology/Invention Resource)

A wise use: Don't spend too much time on computer games.

Chapter 7

Chapter 7 of the *Science on Target for Grade 3 Student Workbook*, covers the National Content Standards for the History & Nature of Science. The standards are as follows:

History and Nature of Science

Science as a human endeavor

- Science and technology have been practiced by people for a long time.
- Men and women have made a variety of contributions throughout the history of Science and Technology.
- Although men and women have learned much about objects, events, and phenomena in nature, more remains to be understood. Science will never be finished.
- Many people chose Science as a career.

All of the pages from Chapter 7 of the *Science on Target for Grade 3 Student Workbook*, are reproduced in this Parent/Teacher Edition in reduced-page format with sample answers. These activities will help your students develop the skills necessary to understand the History and Nature of Science.

Students should use the "Clues for Success" Checklist, for each activity in this section, as a tool to help them do their best work.

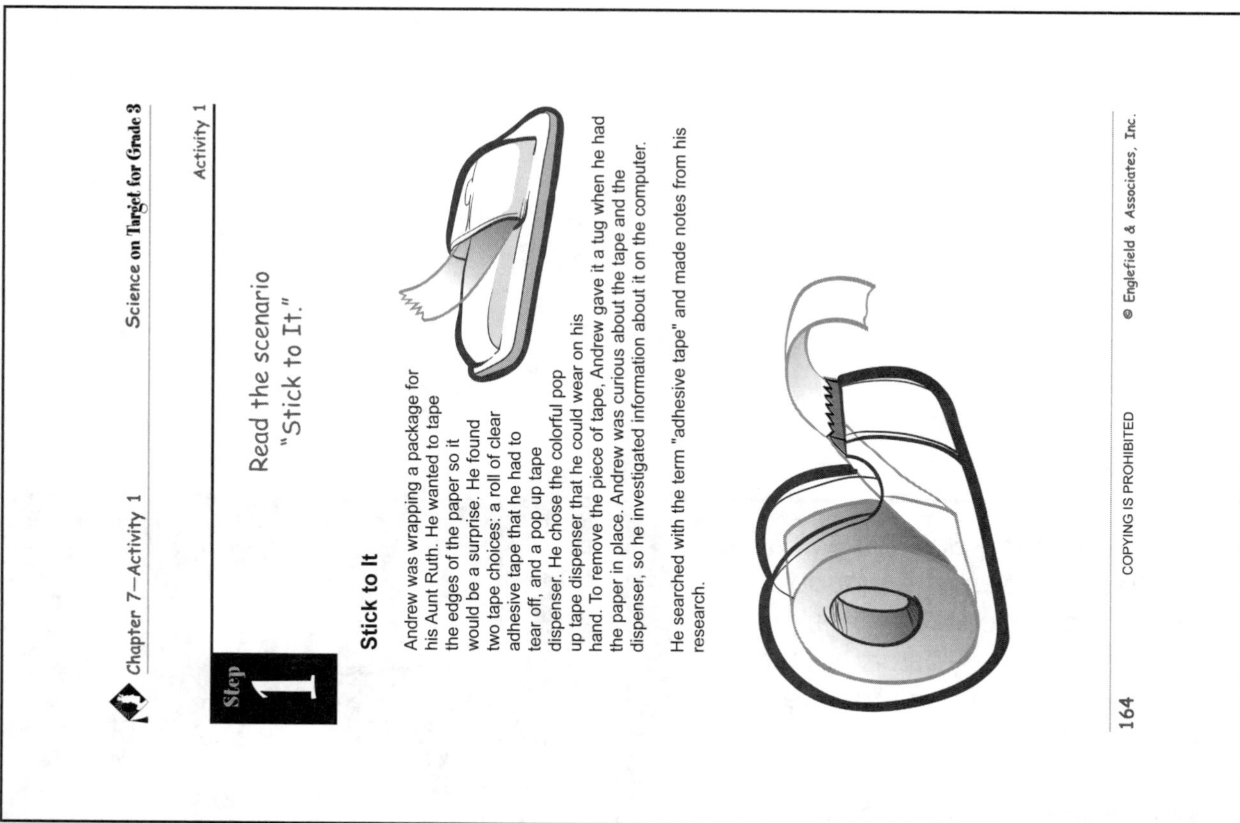
Chapter 7—Activity 1 Science on Target for Grade 3

Activity 1

Step 1

Read the scenario "Stick to It."

Stick to It

Andrew was wrapping a package for his Aunt Ruth. He wanted to tape the edges of the paper so it would be a surprise. He found two tape choices: a roll of clear adhesive tape that he had to tear off, and a pop up tape dispenser. He chose the colorful pop up tape dispenser that he could wear on his hand. To remove the piece of tape, Andrew gave it a tug when he had the paper in place. Andrew was curious about the tape and the dispenser, so he investigated information about it on the computer.

He searched with the term "adhesive tape" and made notes from his research.

164 COPYING IS PROHIBITED © Englefield & Associates, Inc.

Note: Student answers may vary. Example responses are for use as a guide.

Chapter 7

History & Nature of Science

The activities in this section of the book will focus on the History and Nature of Science.

The activities in this section will help you learn that:

- science is created and conducted by people,
- science has been practiced by people for a long time,
- men and women choose careers in Science and make contributions in the areas of Science and Technology,
- scientists are always adding to what people are learning about the world, and
- science will never be finished.

Use the "Clues for Success" Checklist, as you complete each activity in this section, as a tool to help you do your best work.

© Englefield & Associates, Inc. COPYING IS PROHIBITED 163

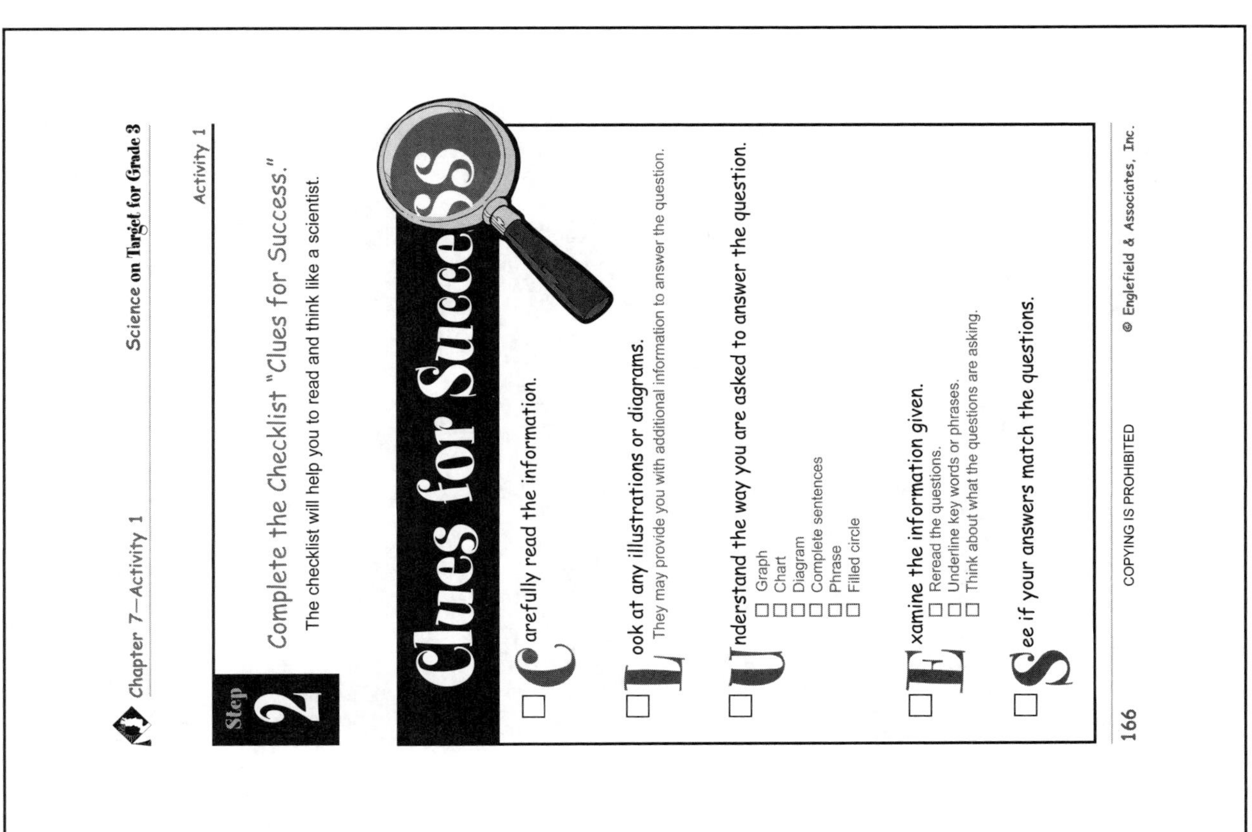

Chapter 7—Activity 1 Science on Target for Grade 3

Step 2

Activity 1

Complete the Checklist "Clues for Success."

The checklist will help you to read and think like a scientist.

Clues for Success

- ☐ **C**arefully read the information.
- ☐ **L**ook at any illustrations or diagrams.
  They may provide you with additional information to answer the question.
- ☐ **U**nderstand the way you are asked to answer the question.
  - ☐ Graph
  - ☐ Chart
  - ☐ Diagram
  - ☐ Complete sentences
  - ☐ Phrase
  - ☐ Filled circle
- ☐ **E**xamine the information given.
  - ☐ Reread the questions.
  - ☐ Underline key words or phrases.
  - ☐ Think about what the questions are asking.
- ☐ **S**ee if your answers match the questions.

166 COPYING IS PROHIBITED © Englefield & Associates, Inc.

Note: Student answers may vary. Example responses are for use as a guide.

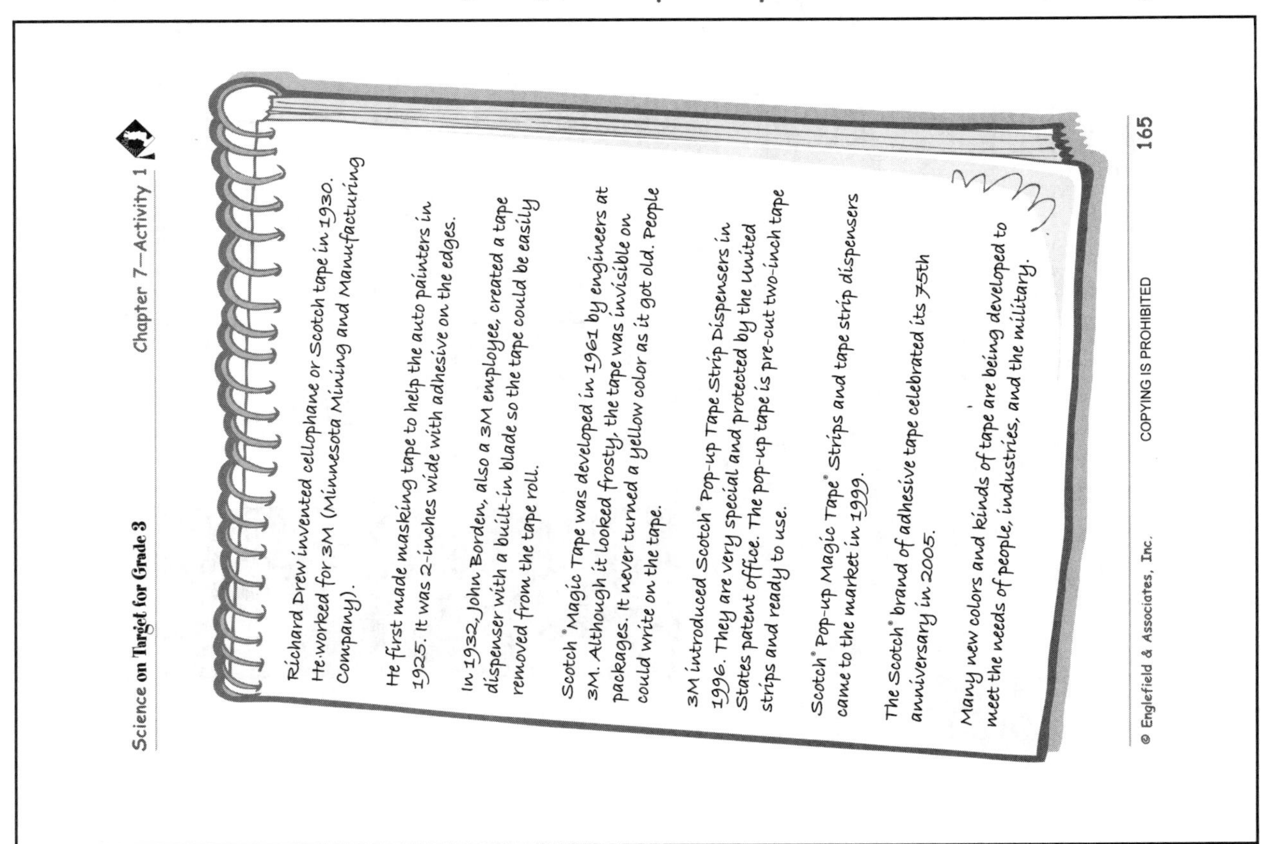

Science on Target for Grade 3 Chapter 7—Activity 1

Richard Drew invented cellophane or Scotch tape in 1930. He worked for 3M (Minnesota Mining and Manufacturing Company).

He first made masking tape to help the auto painters in 1925. It was 2-inches wide with adhesive on the edges.

In 1932, John Borden, also a 3M employee, created a tape dispenser with a built-in blade so the tape could be easily removed from the tape roll.

Scotch® Magic Tape was developed in 1961 by engineers at 3M. Although it looked frosty, the tape was invisible on packages. It never turned a yellow color as it got old. People could write on the tape.

3M introduced Scotch® Pop-up Tape Strip Dispensers in 1996. They are very special and protected by the United States patent office. The pop-up tape is pre-cut two-inch tape strips and ready to use.

Scotch® Pop-up Magic Tape® Strips and tape strip dispensers came to the market in 1999.

The Scotch® brand of adhesive tape celebrated its 75th anniversary in 2005.

Many new colors and kinds of tape are being developed to meet the needs of people, industries, and the military.

© Englefield & Associates, Inc. COPYING IS PROHIBITED 165

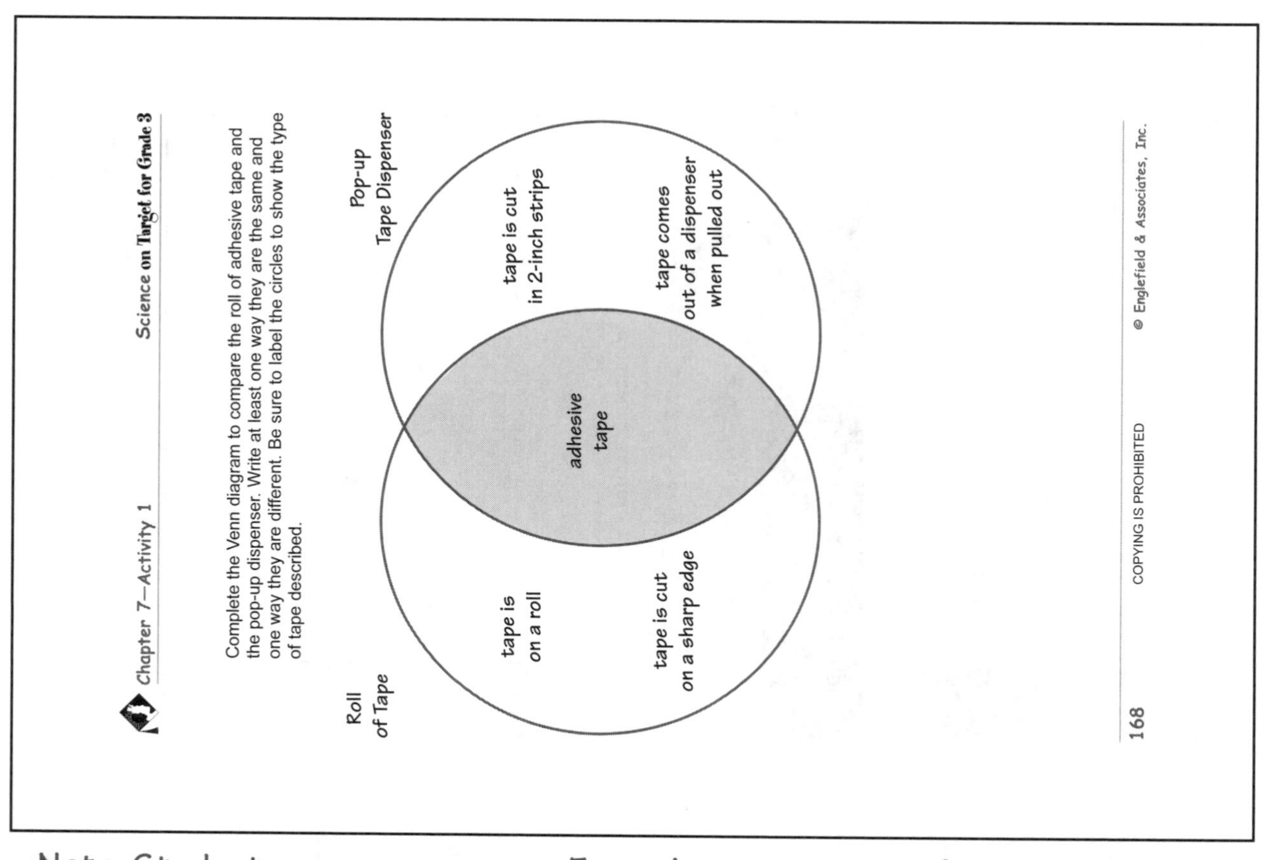

Chapter 7—Activity 1 Science on Target for Grade 3

Complete the Venn diagram to compare the roll of adhesive tape and the pop-up dispenser. Write at least one way they are the same and one way they are different. Be sure to label the circles to show the type of tape described.

168 COPYING IS PROHIBITED © Englefield & Associates, Inc.

Note: Student answers may vary. Example responses are for use as a guide.

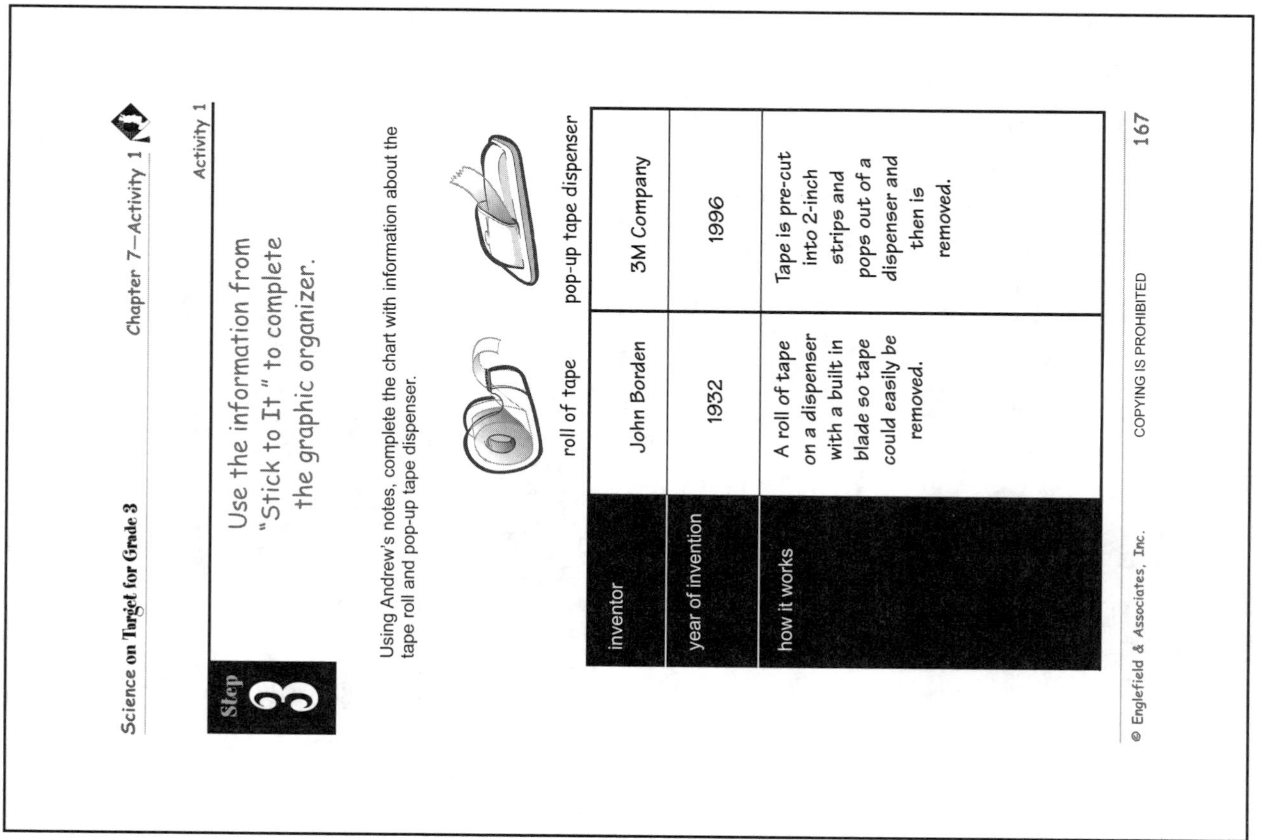

Science on Target for Grade 3 Chapter 7—Activity 1

Step 3 Activity 1

Use the information from "Stick to It" to complete the graphic organizer.

Using Andrew's notes, complete the chart with information about the tape roll and pop-up tape dispenser.

	roll of tape	pop-up tape dispenser
inventor	John Borden	3M Company
year of invention	1932	1996
how it works	A roll of tape on a dispenser with a built in blade so tape could easily be removed.	Tape is pre-cut into 2-inch strips and pops out of a dispenser and then is removed.

© Englefield & Associates, Inc. COPYING IS PROHIBITED 167

Activity 1

Step 4 Answer the following questions for "Stick to It" using information from your graphic organizer.

1. If you had the choice would you use the roll of tape dispenser or the pop-up tape dispenser? Use a complete sentence to give a reason for your choice.

   ● roll of tape ○ pop up tape dispenser

   Reason: I would use a roll of tape because I could use less than 2 inches of tape at a time.

2. Science and inventions help people. Andrew thought the pop-up tape dispenser would help the class when they were wrapping surprise boxes for the school carnival.

   List **two** ways the pop-up tape dispenser might help the students.

   1. They could wrap the boxes faster when the tape pops out of the dispenser.
   2. All of the boxes would have the same amount of tape used in wrapping them.

Note: Student answers may vary. Example responses are for use as a guide.

3. Science and inventions are never finished. How does learning about the history of adhesive tape show this? Use a complete sentence to answer the question.

   The tape was changed from the original tape used to help auto painters to a tape used in everyday life. Technology is making the tape easier to use today.

Chapter 7—Activity 2   Science on Target for Grade 3

172   COPYING IS PROHIBITED   © Englefield & Associates, Inc.

Note: Student answers may vary. Example responses are for use as a guide.

Science on Target for Grade 3   Chapter 7—Activity 2

Activity 2

Step 1

Read the scenario "Scientists and Their Contributions."

Scientists and Their Contributions

The teacher wrote the following assignment on the board for the students to complete after recess.

- Research a scientist or inventor.
- Create a card with the person's picture.
- Under the picture, write the country where the person worked.
- Write a complete sentence about the individual's contribution.

The students on the "Explorer Team" created these cards:

© Englefield & Associates, Inc.   COPYING IS PROHIBITED   171

Step 2

Complete the Checklist "Clues for Success."

The checklist will help you to read and think like a scientist.

Activity 2

Clues for Success

- [ ] **C**arefully read the information.
- [ ] **L**ook at any illustrations or diagrams.
  They may provide you with additional information to answer the question.
- [ ] **U**nderstand the way you are asked to answer the question.
  - [ ] Graph
  - [ ] Chart
  - [ ] Diagram
  - [ ] Complete sentences
  - [ ] Phrase
  - [ ] Filled circle
- [ ] **E**xamine the information given.
  - [ ] Reread the questions.
  - [ ] Underline key words or phrases.
  - [ ] Think about what the questions are asking.
- [ ] **S**ee if your answers match the questions.

Note: Student answers may vary. Example responses are for use as a guide.

Activity 2

Step 4

Answer the following questions for "Scientists and Their Contributions" using information from your graphic organizer.

Men and women make many scientific discoveries and inventions. Which scientist or inventor named in "Scientists and Their Contributions" would make the following statements? Write the name of the person on the box next to the statement.

Statement	Person
Traffic is stopped at the intersections.	*Garrett Morgan*
There is a lot of interesting things to learn about the sea.	*Jacques Cousteau*
I want to go to the moon.	*Sally Ride*
I must practice safety in my chemistry lab.	*Marie Curie*
My inventions will help people read at night.	*Thomas Edison*
Underwater exploration is fun.	*Jacques Cousteau*
Being the first women chosen to do something is an honor!	*Sally Ride*

Note: Student answers may vary. Example responses are for use as a guide.

Activity 2

Step 3

Use the information from "Scientists and Their Contributions" to complete the graphic organizer.

Complete the chart below with the information found on the Explorer Team's cards.

	true	false	evidence
All scientists are from France.	○	●	*Marie Curie was from Poland.*
Men are scientists.	●	○	*Thomas Edison, Jacques Cousteau, and Garrett Morgan are scientists.*
Women can be explorers.	●	○	*Sally Ride is a space explorer.*
All exploration is in space.	○	●	*Exploration can be underwater and in a laboratory.*
Scientists can only work in laboratories.	○	●	*Scientists work in space and underwater.*
Some inventions are used in people's homes.	●	○	*The light bulb is used in people's homes.*

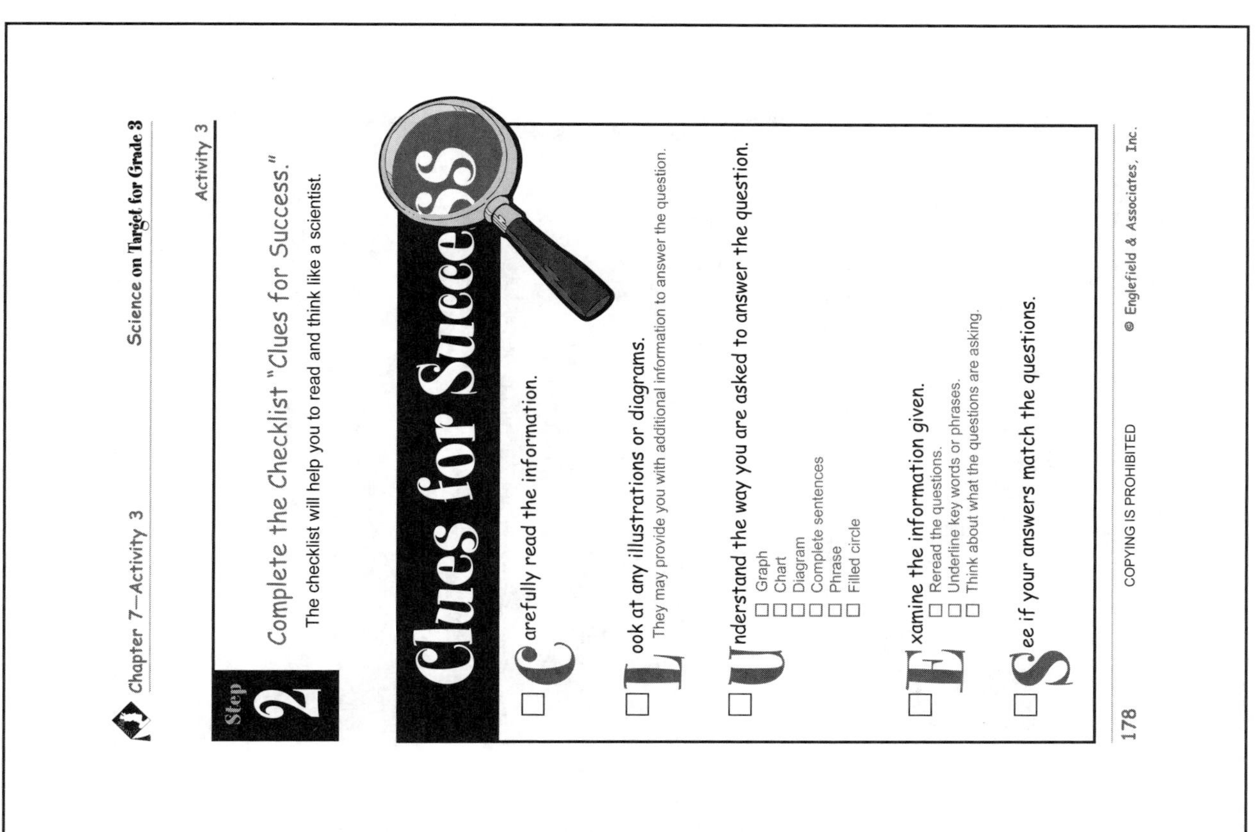

Chapter 7—Activity 3 Science on Target for Grade 3

Activity 3

Step 2 Complete the Checklist "Clues for Success."

The checklist will help you to read and think like a scientist.

Clues for Success

- ☐ **C**arefully read the information.
- ☐ **L**ook at any illustrations or diagrams.
  They may provide you with additional information to answer the question.
- ☐ **U**nderstand the way you are asked to answer the question.
  - ☐ Graph
  - ☐ Chart
  - ☐ Diagram
  - ☐ Complete sentences
  - ☐ Phrase
  - ☐ Filled circle
- ☐ **E**xamine the information given.
  - ☐ Reread the questions.
  - ☐ Underline key words or phrases.
  - ☐ Think about what the questions are asking.
- ☐ **S**ee if your answers match the questions.

178 COPYING IS PROHIBITED © Englefield & Associates, Inc.

Note: Student answers may vary. Example responses are for use as a guide.

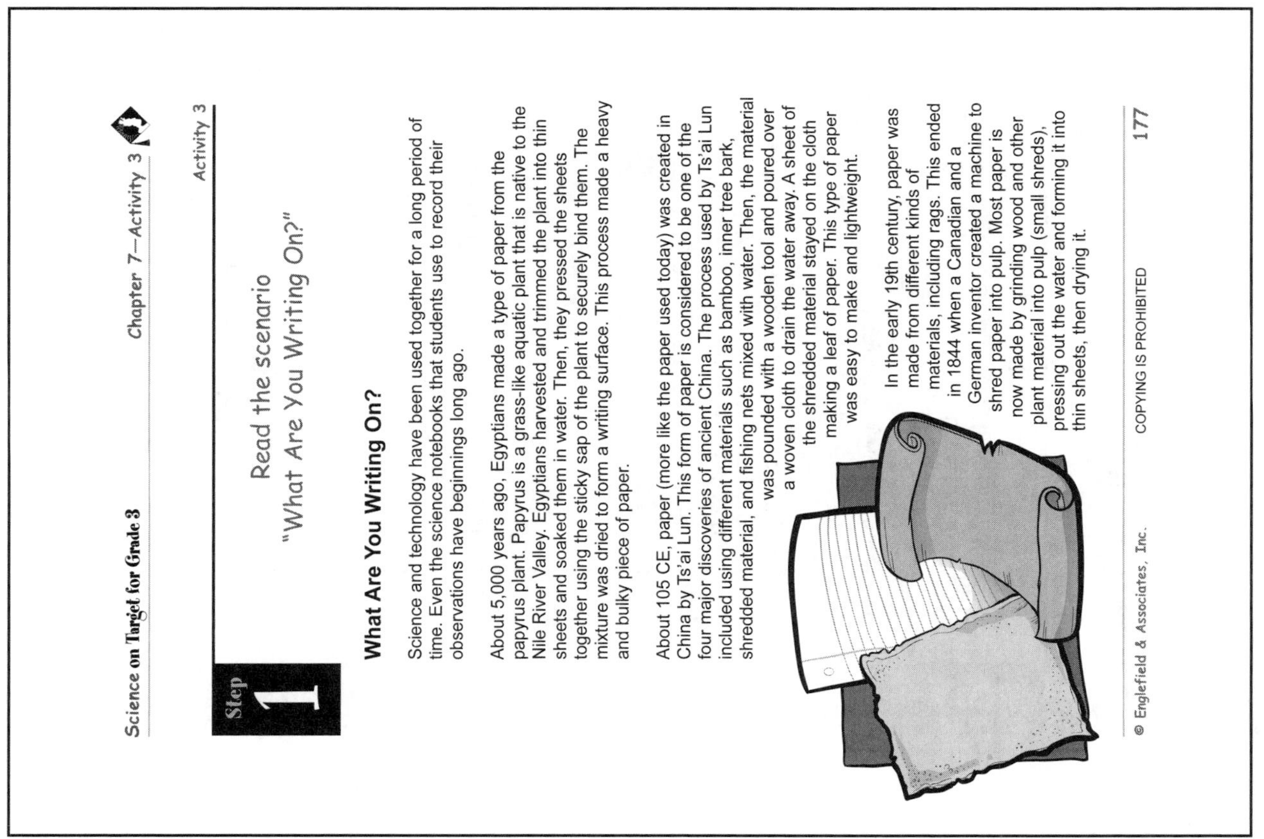

Science on Target for Grade 3 Chapter 7—Activity 3

Activity 3

Step 1 Read the scenario "What Are You Writing On?"

What Are You Writing On?

Science and technology have been used together for a long period of time. Even the science notebooks that students use to record their observations have beginnings long ago.

About 5,000 years ago, Egyptians made a type of paper from the papyrus plant. Papyrus is a grass-like aquatic plant that is native to the Nile River Valley. Egyptians harvested and trimmed the plant into thin sheets and soaked them in water. Then, they pressed the sheets together using the sticky sap of the plant to securely bind them. The mixture was dried to form a writing surface. This process made a heavy and bulky piece of paper.

About 105 CE, paper (more like the paper used today) was created in China by Ts'ai Lun. This form of paper is considered to be one of the four major discoveries of ancient China. The process used by Ts'ai Lun included using different materials such as bamboo, inner tree bark, shredded material, and fishing nets mixed with water. Then, the material was pounded with a wooden tool and poured over a woven cloth to drain the water away. A sheet of the shredded material stayed on the cloth making a leaf of paper. This type of paper was easy to make and lightweight.

In the early 19th century, paper was made from different kinds of materials, including rags. This ended in 1844 when a Canadian and a German inventor created a machine to shred paper into pulp. Most paper is now made by grinding wood and other plant material into pulp (small shreds), pressing out the water and forming it into thin sheets, then drying it.

© Englefield & Associates, Inc. COPYING IS PROHIBITED 177

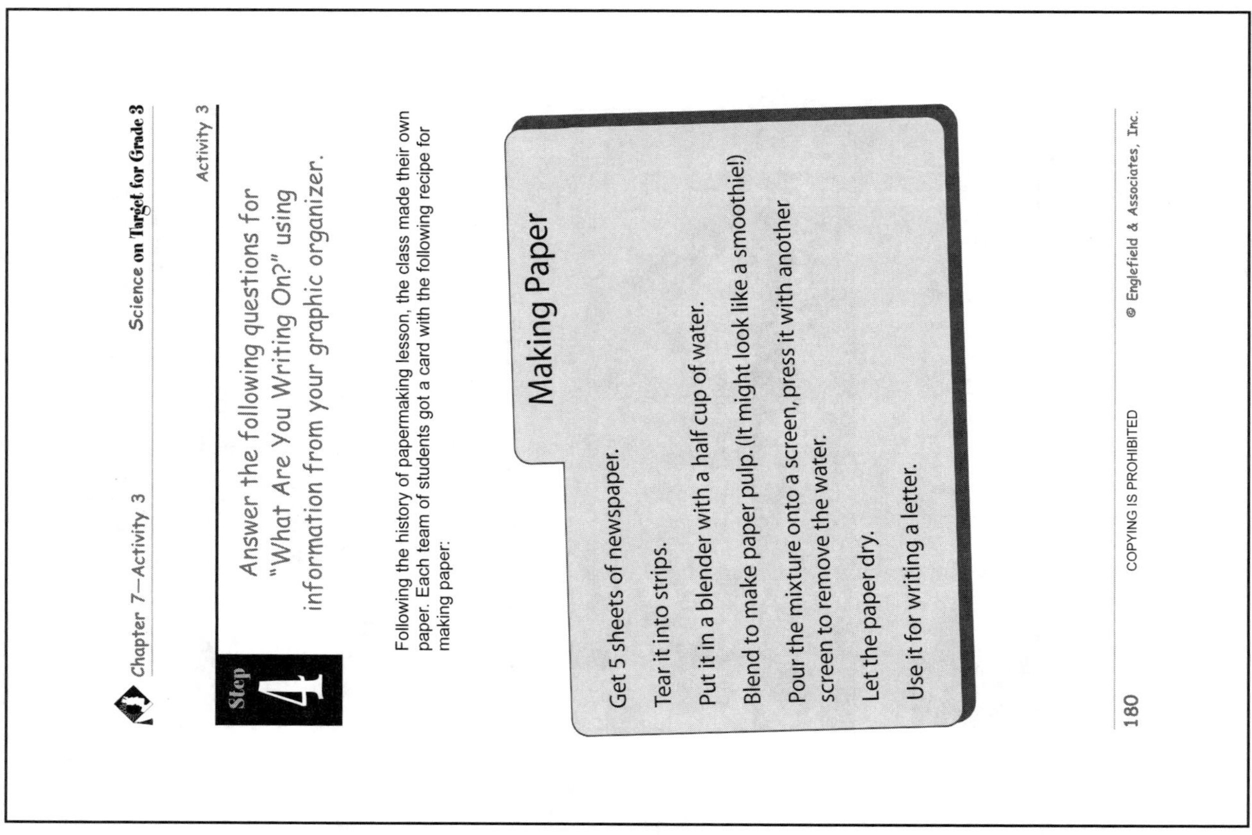

Chapter 7—Activity 3 Science on Target for Grade 3

Activity 3

Step 4

Answer the following questions for "What Are You Writing On?" using information from your graphic organizer.

Following the history of papermaking lesson, the class made their own paper. Each team of students got a card with the following recipe for making paper:

Making Paper

Get 5 sheets of newspaper.

Tear it into strips.

Put it in a blender with a half cup of water.

Blend to make paper pulp. (It might look like a smoothie!)

Pour the mixture onto a screen, press it with another screen to remove the water.

Let the paper dry.

Use it for writing a letter.

180 COPYING IS PROHIBITED © Englefield & Associates, Inc.

Note: Student answers may vary. Example responses are for use as a guide.

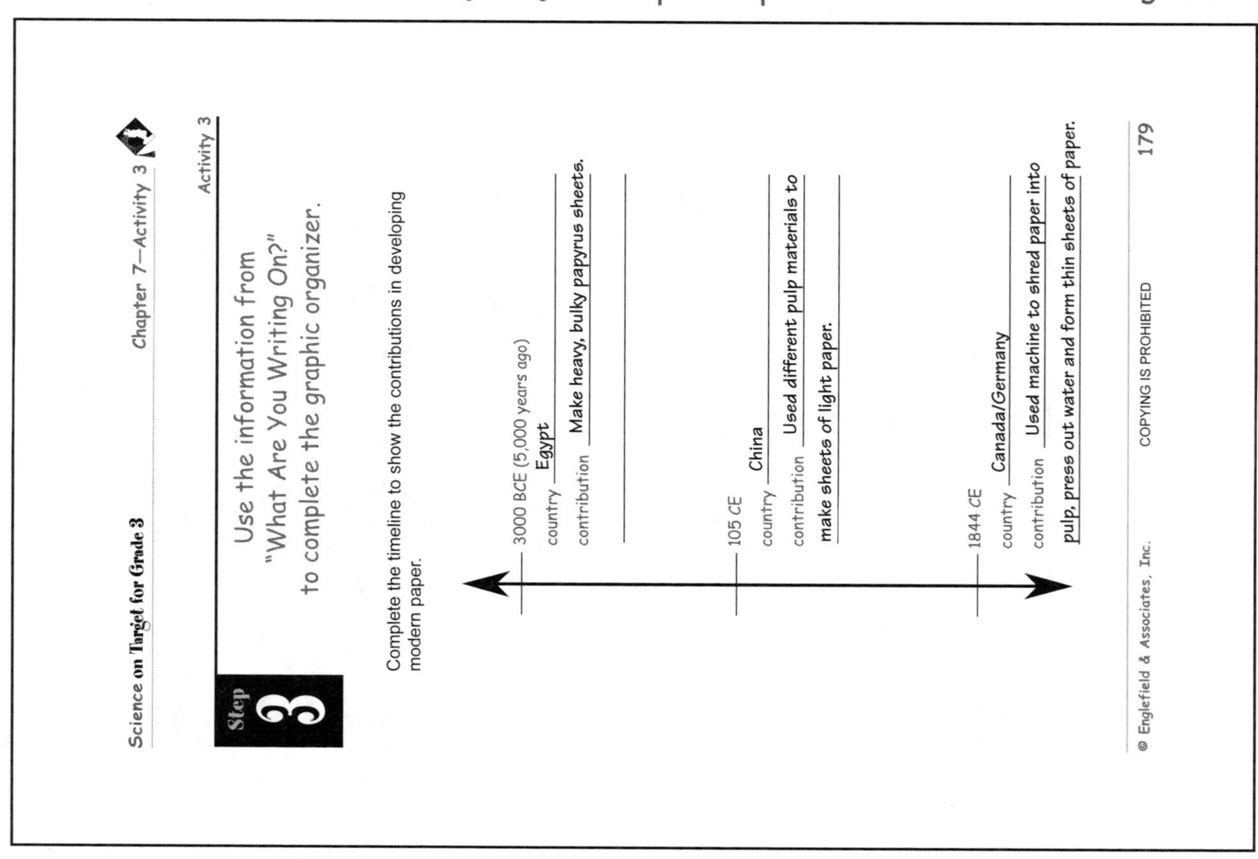

Science on Target for Grade 3 Chapter 7—Activity 3

Activity 3

Step 3

Use the information from "What Are You Writing On?" to complete the graphic organizer.

Complete the timeline to show the contributions in developing modern paper.

3000 BCE (5,000 years ago)
country Egypt
contribution Make heavy, bulky papyrus sheets.

105 CE
country China
contribution Used different pulp materials to make sheets of light paper.

1844 CE
country Canada/Germany
contribution Used machine to shred paper into pulp, press out water and form thin sheets of paper.

© Englefield & Associates, Inc. COPYING IS PROHIBITED 179

1. List **two** ways the paper making process used by the class is the same as the process used by Ts'ai Lun.

   1. They pressed out the water.

   2. They let the paper sheet dry.

2. List **two** ways the paper making process used by the class is different from the process used by Ts'ai Lun.

   1. The class used newspaper to make pulp.

   2. The class used a blender to shred the newspaper.

Note: Student answers may vary. Example responses are for use as a guide.

Activity 4

Step 1

Read the scenario "Polio Affects a President and Children."

Polio Affects a President and Children

The third grade students were surprised to learn that Franklin Roosevelt had polio and used a wheel chair during his terms as president of the United States. They were interested in learning more about polio and researched the illness during their computer class.

Here is some of the information they learned.

During the 1940's and 1950's, Poliomyelitis (polio) was an epidemic disease that affected children more than adults. It spread quickly through the population. Polio caused difficulty breathing, paralysis (can't move your arms and legs), or even death. Before the first polio vaccine was developed in the 1950s, thousands of children got polio every year.

At the time, a child with polio was placed into a large metal tube called an "iron lung" to help the child breath. Today, the "iron lung" has largely been replaced by a ventilator (a smaller, often hand-held tool). Children with polio generally could not move by themselves. Today, children whose legs are paralyzed use crutches, special braces, or wheelchairs to help them move around.

Doctors can help prevent children from getting polio by giving them a vaccine. A vaccine helps produce protection to fight against getting the disease. Two main scientists helped wipe out polio by creating vaccines. The first polio vaccine developed by Jonas Salk in the 1950's was given by drops into the mouth (orally). Later, Albert Sabin developed a vaccine that was given by injection (a "shot").

POLIO VACCINE

Today, there is still no cure for polio, but the polio vaccine (given about 4 times from birth–6 years old) almost always protects a child for life. Fortunately, the use of the polio vaccine has made the disease very rare in most parts of the world.

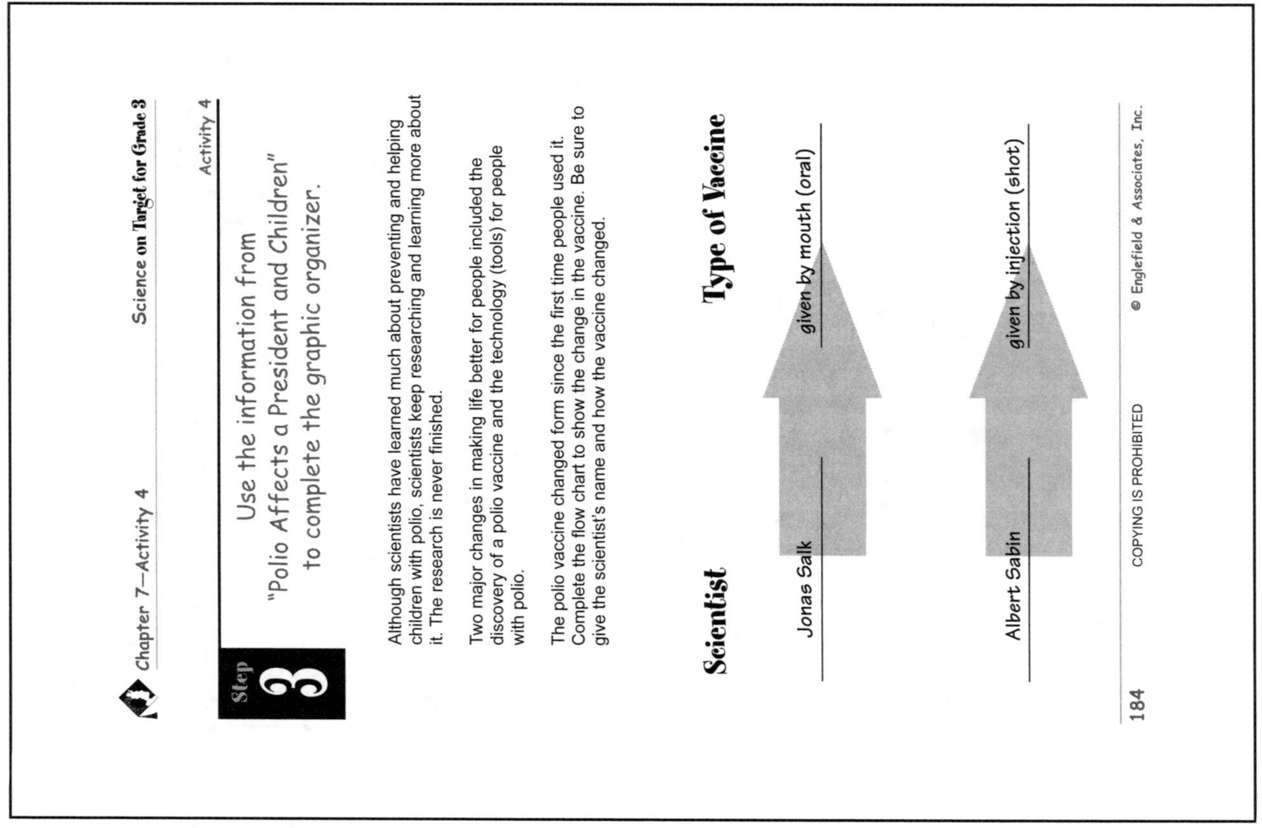

Step 3 Use the information from "Polio Affects a President and Children" to complete the graphic organizer.

Although scientists have learned much about preventing and helping children with polio, scientists keep researching and learning more about it. The research is never finished.

Two major changes in making life better for people included the discovery of a polio vaccine and the technology (tools) for people with polio.

The polio vaccine changed form since the first time people used it. Complete the flow chart to show the change in the vaccine. Be sure to give the scientist's name and how the vaccine changed.

Scientist	Type of Vaccine
Jonas Salk	given by mouth (oral)
Albert Sabin	given by injection (shot)

Note: Student answers may vary. Example responses are for use as a guide.

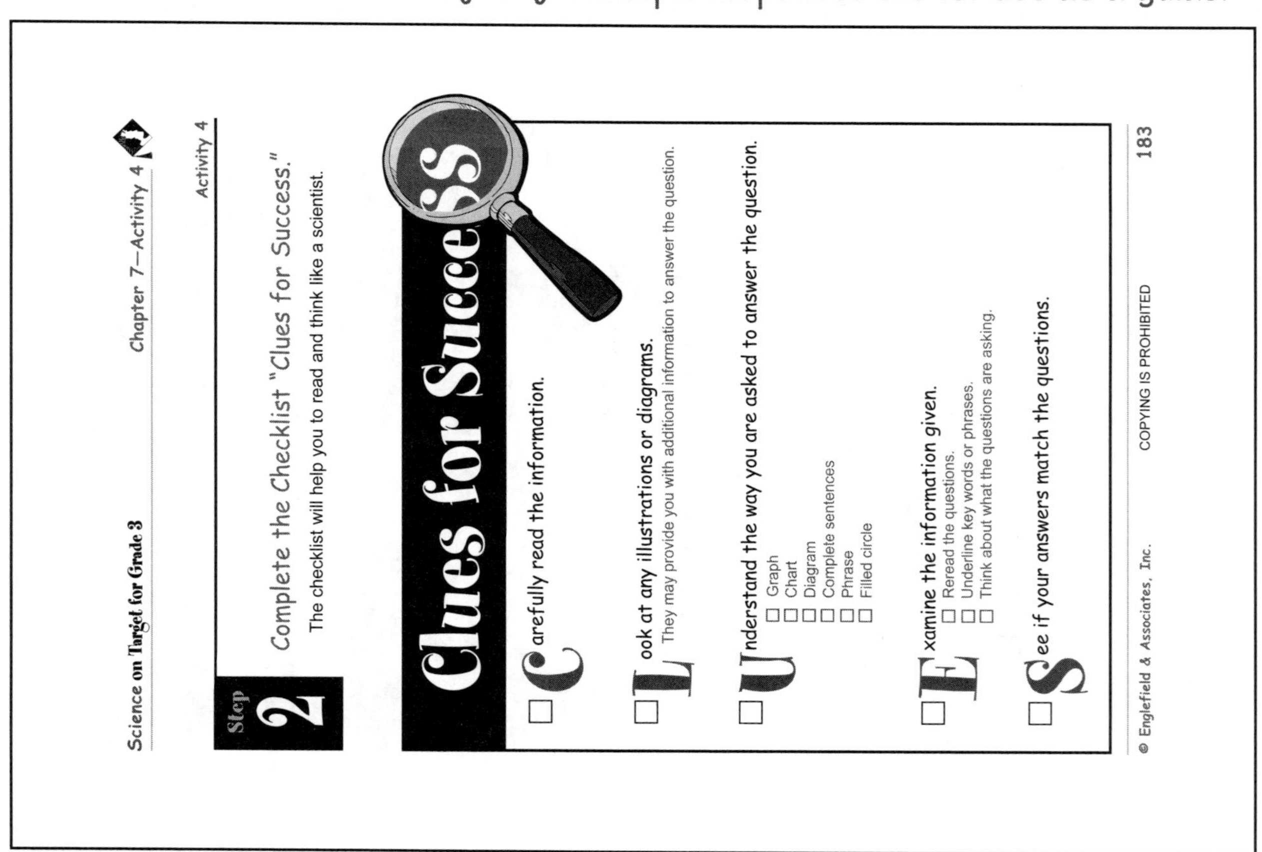

Step 2 Complete the Checklist "Clues for Success."

The checklist will help you to read and think like a scientist.

Clues for Success

- ☐ **C**arefully read the information.
- ☐ **L**ook at any illustrations or diagrams.
  They may provide you with additional information to answer the question.
- ☐ **U**nderstand the way you are asked to answer the question.
  - ☐ Graph
  - ☐ Chart
  - ☐ Diagram
  - ☐ Complete sentences
  - ☐ Phrase
  - ☐ Filled circle
- ☐ **E**xamine the information given.
  - ☐ Reread the questions.
  - ☐ Underline key words or phrases.
  - ☐ Think about what the questions are asking.
- ☐ **S**ee if your answers match the questions.

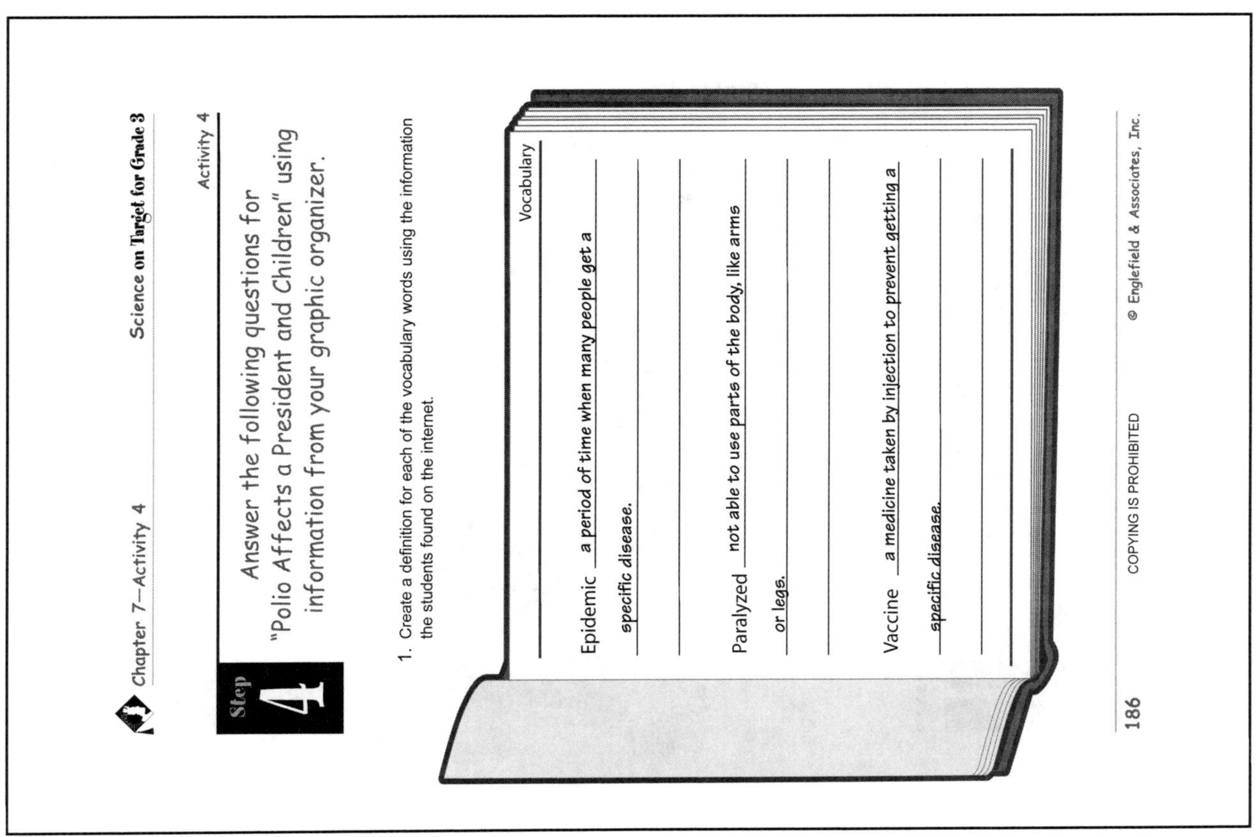

Activity 4

Step 4

Answer the following questions for "Polio Affects a President and Children" using information from your graphic organizer.

1. Create a definition for each of the vocabulary words using the information the students found on the internet.

Vocabulary

Epidemic *a period of time when many people get a specific disease.*

Paralyzed *not able to use parts of the body, like arms or legs.*

Vaccine *a medicine taken by injection to prevent getting a specific disease.*

Note: Student answers may vary. Example responses are for use as a guide.

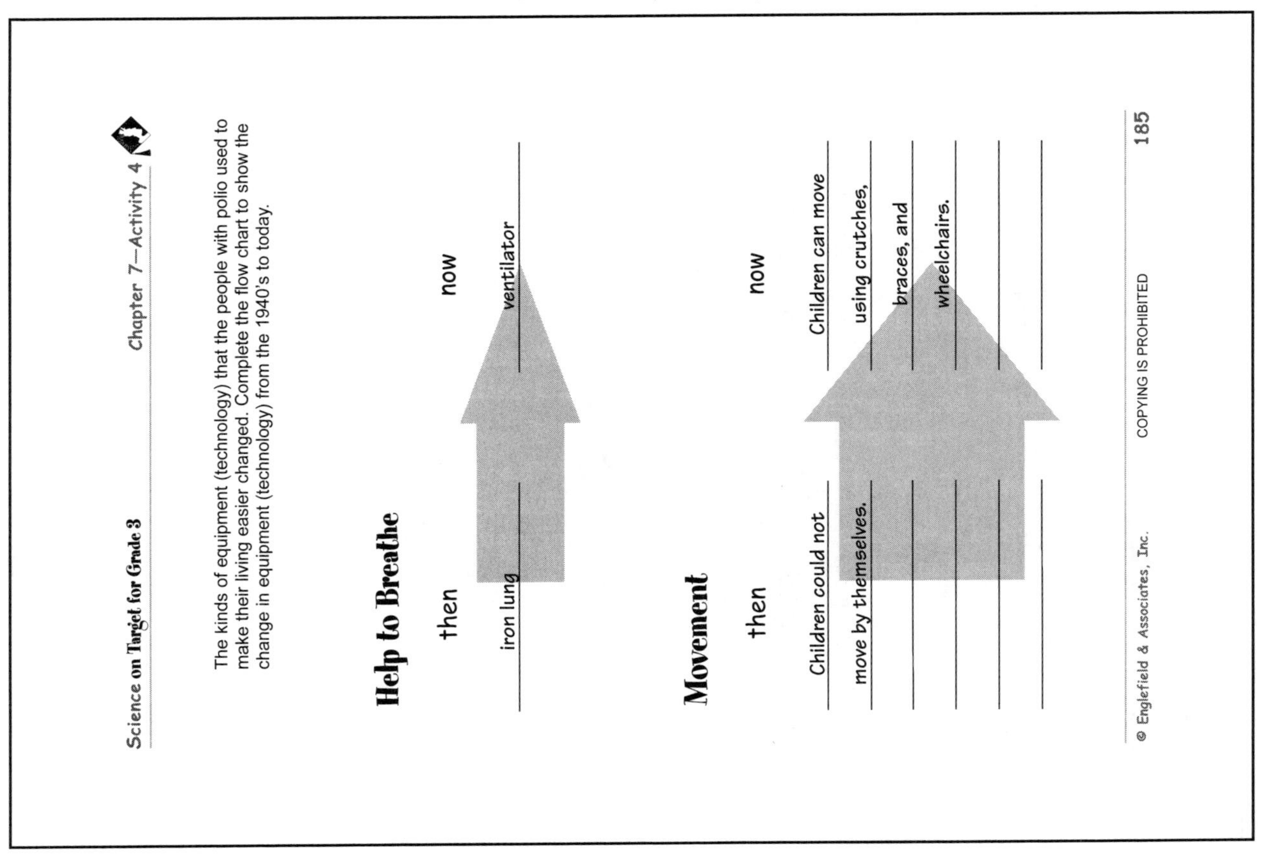

The kinds of equipment (technology) that the people with polio used to make their living easier changed. Complete the flow chart to show the change in equipment (technology) from the 1940's to today.

Help to Breathe

then	now
iron lung	*ventilator*

Movement

then	now
Children could not move by themselves.	*Children can move using crutches, braces, and wheelchairs.*

Step 1

Read the scenario "Scientists at the Zoo."

Scientists at the Zoo

On the field trip to the zoo, the students met scientists that worked in different kinds of jobs at the zoo. They took pictures of the scientists to create a new bulletin board about the zoo careers.

Note: Student answers may vary. Example responses are for use as a guide.

2. Although polio is very rare in the world today, do you think it is important for scientists to keep learning more about preventing polio? Use a completely filled circle to show your answer choice. Then, use a complete sentence to explain why.

● yes ○ no

Why?

Scientists should keep studying polio to make sure the polio epidemic never returns, and maybe discover a cure.

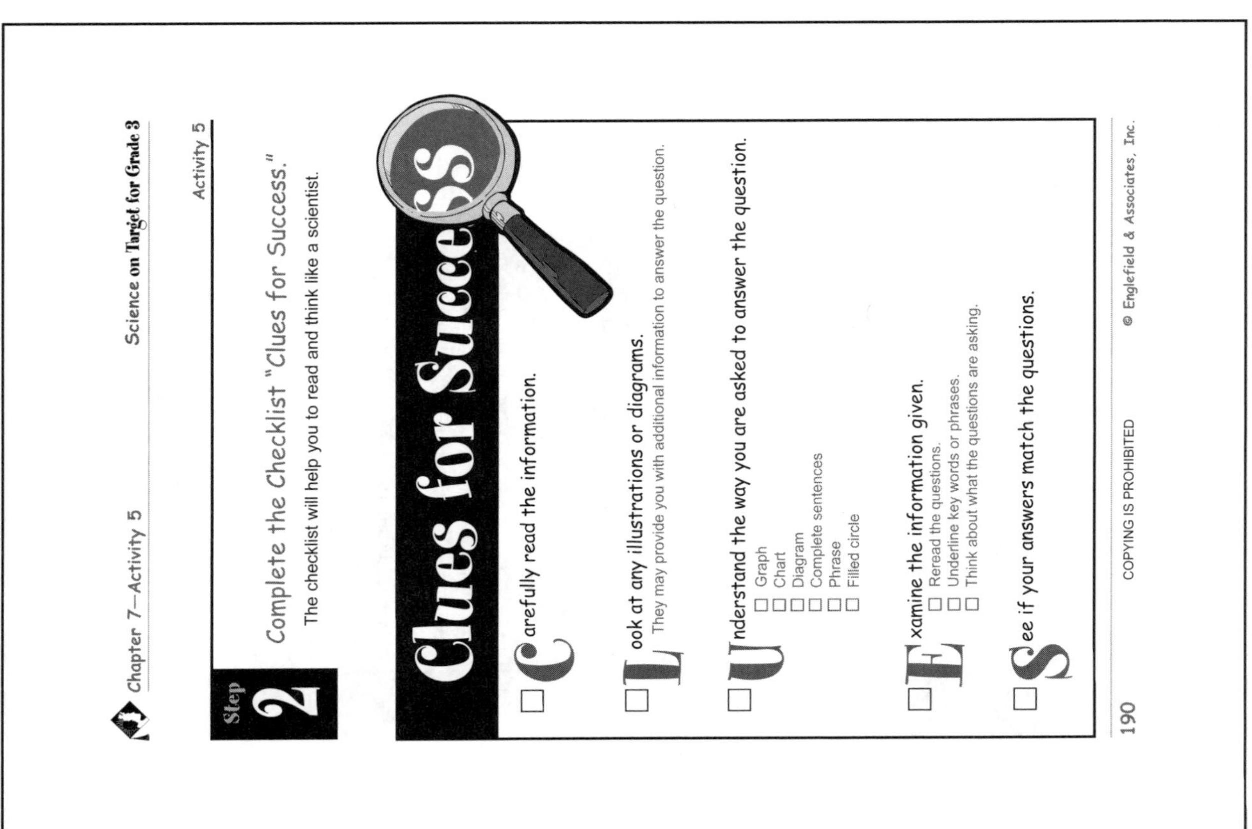

Activity 5

Step 2

Complete the Checklist "Clues for Success."

The checklist will help you to read and think like a scientist.

Clues for Success

☐ **C**arefully read the information.

☐ **L**ook at any illustrations or diagrams.
They may provide you with additional information to answer the question.

☐ **U**nderstand the way you are asked to answer the question.
- ☐ Graph
- ☐ Chart
- ☐ Diagram
- ☐ Complete sentences
- ☐ Phrase
- ☐ Filled circle

☐ **E**xamine the information given.
- ☐ Reread the questions.
- ☐ Underline key words or phrases.
- ☐ Think about what the questions are asking.

☐ **S**ee if your answers match the questions.

Note: Student answers may vary. Example responses are for use as a guide.

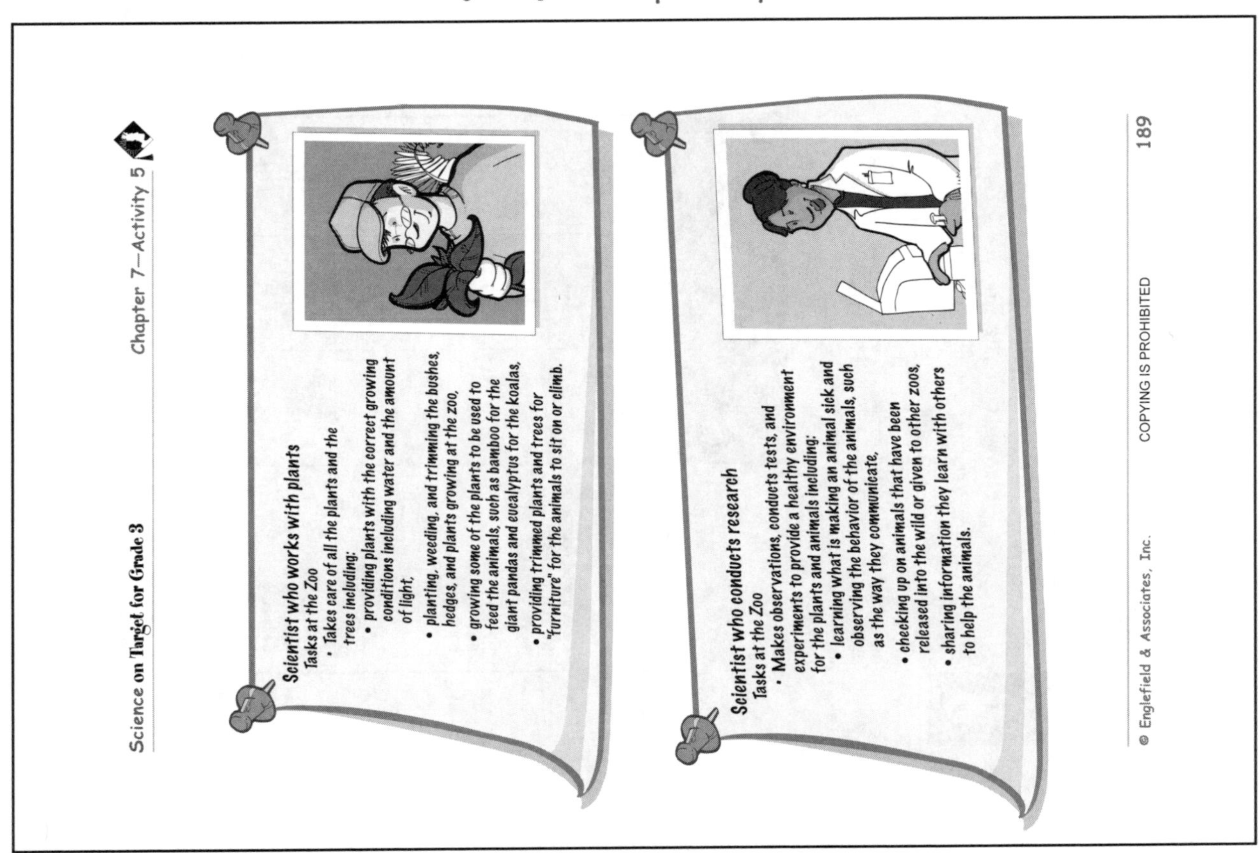

Scientist who works with plants
Tasks at the Zoo
- Takes care of all the plants and the trees including:
  - providing plants with the correct growing conditions including water and the amount of light,
  - planting, weeding, and trimming the bushes, hedges, and plants growing at the zoo,
  - growing some of the plants to be used to feed the animals, such as bamboo for the giant pandas and eucalyptus for the koalas,
  - providing trimmed plants and trees for "furniture" for the animals to sit on or climb.

Scientist who conducts research
Tasks at the Zoo
- Makes observations, conducts tests, and experiments to provide a healthy environment for the plants and animals including:
  - learning what is making an animal sick and observing the behavior of the animals, such as the way they communicate,
  - checking up on animals that have been released into the wild or given to other zoos,
  - sharing information they learn with others to help the animals.

Activity 5

Step 3

Use the information from "Scientists at the Zoo" to complete the graphic organizer.

The students made a list of jobs they saw people doing at the zoo. Their teacher created a chart and asked the students to place an **X** to show the kind of career the person had.

	Working with Animals	Working with Plants	Animal Research
Provided food for the penguins	X		
Grew the bamboo to feed the pandas		X	
Cut the trees to make shelters in the cages		X	
Talked to the students about the food the manatees need	X		
Worked to discover what was making the birds sick	X		X
Made a list of how the gorillas communicated	X		X

Note: Student answers may vary. Example responses are for use as a guide.

Activity 5

Step 4

Answer the following questions for "Scientists at the Zoo" using information from your graphic organizer.

Chip said that he saw other people working at the zoo. He made a list of some of the people he saw working and he said they used science in their jobs too. The teacher agreed with Chip. Read the description of the worker and use a complete sentence to show how science is involved in his or her job.

A diver was cleaning the aquarium. How is science involved in this job?

The diver is using tools and knowledge to provide a good habitat for the organisms in the aquarium.

The construction workers were building a new reptile building. How is science involved in this job?

Construction workers learned about the environmental needs of the reptiles, the zoo grounds, and visitors to build a good structure.

A local animal shelter presented a program with animals they saved. How is science involved in this job?

The workers at the animal shelter learned about the medical needs of the animals and how to help them recover from their abuse.